ALL OF PHYSICS (ALMOST) IN

ALL OF PHYSICS (ALMOST) IN

Bruno Mansoulié

illustrations by Lison Bernet

World Scientific

NEW JERSEY • LONDON • SINGAPORE • BEIJING • SHANGHAI • HONG KONG • TAIPEI • CHENNAI • TOKYO

Published by

World Scientific Publishing Co. Pte. Ltd.
5 Toh Tuck Link, Singapore 596224
USA office: 27 Warren Street, Suite 401-402, Hackensack, NJ 07601
UK office: 57 Shelton Street, Covent Garden, London WC2H 9HE

British Library Cataloguing-in-Publication Data
A catalogue record for this book is available from the British Library.

ALL OF PHYSICS (ALMOST) IN 15 EQUATIONS

Illustrations by Lison Bernet
Solstice / Lison Bernet © Flammarion

The work was originally published in French by Flammarion as
Toute la physique en 15 équations (ou presque...)
This edition is published by World Scientific Publishing Company Pte Ltd
by arrangement with Flammarion, Paris, France.

ISBN 978-981-3273-40-5 (pbk)

For any available supplementary material, please visit
https://www.worldscientific.com/worldscibooks/10.1142/11075#t=suppl

Typeset by Stallion Press
Email: enquiries@stallionpress.com

Printed in Singapore

Foreword

I am a scientist in the field of Particle Physics, which should rather be called "Physics of the Elementary". My goal is to understand the fundamental constituents of Nature and their interactions, from the smallest particles to the largest structures of the Universe.

With this book, I would like to introduce all the readers with the equations of physics, if possible. I have in mind those who never encountered them, or of those for whom the first encounters have been disappointing, or even traumatic.

We, physicists, live with two visions of the world: on the one hand the same vision as everybody's, and on the other hand the vision provided by the laws of physics: the equations. The general public easily admits this duality for a musician: a musician is an artist, but (s)he also masters a wide theoretical knowledge and a strong technical training. It is much less accepted for a physicist, who is usually depicted as lost in his/her equations and as being disconnected from reality.

It is true that a large fraction of the scientific knowledge appears in the form of laws, written as equations. But even if this sounds surprising, the equations can become so familiar, that the vision of the world "through the equations" and the usual vision can blend together.

Let me take an example. When I look at a rainbow, I do not explicitly have in mind the laws of the refraction of light. Like everybody else, I admire the beauty of the landscape, underlined by the rainbow. But "behind this", there is an entire set of knowledge and practice about light, its propagation, its physical nature, which silently integrates with my vision of the rainbow. Exactly like a musician enjoys the piece he is listening to without becoming aware of its key, its type of chords and its rhythm structure. Of course, if you suddenly ask her, she will certainly provide an accurate answer on all these points...and this will not damage her sensibility for the piece.

Towards the equation of everything

The choice of the equations shown in this book may appear arbitrary to the professional or amateur physicist. The rationale for my choice is that each equation is evidence of a scientific evolution, and sometimes even a true revolution.

Each one displays a particular understanding of the world or of a phenomenon: light, matter, heat, etc. Formulated when mathematics spread into physics, the first laws remain shy, restricted to one particular domain of application, like the laws of reflection and refraction of light. Then the ambition grows: the goal is to fit into an ever growing framework with a universal aim. Thus Newton understands that the attraction between the Earth and the Moon is identical to the force that makes an apple drop from a tree. Similarly, Maxwell unifies electricity and magnetism, then modern physics and chemistry explain how all chemical elements are built from protons, neutrons and electrons. At the start of the 20th century, relativity and quantum mechanics question the very notions of space, time, and matter, before cosmology finally takes the entire universe as a research subject.

Today, the obsession of the physicists for the elementary drives them in an attempt to gather everything into a single "minimal" theory. As we will see at the end of this book, the modern formulation of a physical theory should ideally be contained in one single equation. This is the reason why it is often said that physicists seek "the equation of everything". Of course, this idea is simplistic, but what would the ultimate equation look like? I do not have the answer, but through the examples in this book, I would like to share with you the taste of this quest for an elegant and powerful formula, which could help us understand the world.

Beautiful equations

The equations of physics may arouse respect, fascination, and fear. Even if one has not undertaken long studies, or even is set against all forms of mathematics, one knows $E = mc^2$, and one vaguely associates it to the immense power of nuclear energy. In the eyes of the public, equations are the symbol of the power of science, but also of its coldness, its inhumanity.

When I give a conference for the general public, this is the first demand from the organizers: "No equations, please!". Yet, I reply that some of them are quite simple, since they do not show more than a simple proportionality relation, or the direct dependence of one quantity on another. The effort to understand them is certainly not larger than the effort to decipher the instruction manual of an Internet connection box. Alas, for the average organizer of general public conferences, or for the average journalist, equations just mean a vanishing audience.

The people in the audience are aware that sciences, and the mathematics that serve them, are at work in all aspects of ordinary life, that an airplane or a smartphone have been designed and built by engineers who have mastered and employed equations. But people prefer to ignore that, and leave these unappealing tools to them.

I believe that this mistrust of mathematics in general, and against mathematics in physical laws have deeper grounds than the simple aversion against formulae. Equations constrain us to a kind of intellectual discipline. They do not say today the contrary of what they said the day before. They compel us to clarify our thinking, to avoid vagueness when we talk to somebody. This discipline can be disturbing: vagueness is sometimes so convenient…

I am not claiming that the equations of physics are "true", I am claiming that they are "sincere". When I write the equations of Newtonian mechanics, which, since the 17th century, have been describing the motion of celestial bodies, artillery shells and merry-go-rounds, I do not claim to explain the whole world, why the Sun shines, why flowers grow, or why I have a headache. But these equations offer me a model, a consistent set of relations between the observed positions of planets or between the forces that I feel in a merry-go-round. Then I am free to apply this model judiciously, to the objects and phenomena which I reckoned as belonging to its validity domain. I could even test the limits of this validity domain, perform measurements, make experiments, or simply explore the consequences of these equations in my mind: do they describe the world at very small or very large distances? Etc.

Liberty equation

I often make the following recommendation to students: "The point is not to know everything. The point is to know what you know, and to know what you don't know".

This is the meaning of a good physics equation: it summarizes what is known, in a certain domain. Its variables define the objects and the concepts which we are talking about, no more, no less.

The key word is "freedom". It may sound paradoxical, but to write an equation to describe the world, is not to reduce our vision of the world, but the exact opposite. It amounts to making the choice of a *certain* vision of a *certain* part of the world. This choice is freely adopted, and often based on consensus. Together, we have defined a common language and we agree that a part, also agreed among us, of the phenomena in the world, may be described by the equation. Now we can move forward, use this model in its domain to heal people or manufacture smartphones, and go discover the world beyond this domain. At least, we no longer run the risk of quibbling over definitions, of taking specific cases for generalities, nor of letting others dictate to us our vision of the world.

So much for the rational part. But equations also hold an emotional aspect. Some of them just for their graphic style, even if one does not understand their mathematical or physical meaning. The sensuous curve of a ∂, the aggressive arrow of a $\vec{p}$, or the ambiguous arrow of a $\overleftrightarrow{d}$ …

Some by the elegance of the concepts they use.

Some by their power and the scope of their consequences.

Some from a more personal viewpoint: who taught it to us, to whom we have taught it, etc. The moment when we saw it for the first time in a class or in a book, and the corresponding period in our private life. In which respect it has changed our vison of the world.

The inner spaces it evokes… The dreams it kindles…

Contents

Foreword v

Chapter 1 The law of light reflection 1

But what is a reflection anyway? 3
By the way, what does "to see" mean? 5
Physics of enlightenment 6

Chapter 2 Snell-Descartes law… and the Principle of Least Action 7

Experimenting 9
Whose law is it anyway? 11
When light loses no time 11
A principle of "least action" 12
A formula to sum everything up? 13

Chapter 3 The fundamental principle of dynamics (Newton's second law) 15

Perfect because motionless (?) 17
From fountains to planets 19
A touch of nostalgia for the lost "γ" 20
A rather dynamic physics 21

Chapter 4 The law of gravitation (Newton's first law) 23

The unloved equation 25
Too simple a law? 27
… to conceal ignorance? 28
E pur, si muove! 28
Cosmic chaos 29
A planetary pool game 30

Chapter 5 The ideal gas law 33

An ideal air 35
Good fortune 37

	The air of the law?	37
	An amazing hypothesis	38
	Measuring the size of atoms by blowing air	39
	A revolution in progress	40
Chapter 6	**Hooke's law**	**41**
	The physics of the industrial revolution	43
	Elastic like a metal	45
	The "mechanical power of fire"	46
	Heat, movement, energy	46
Chapter 7	**Navier-Stokes equation**	**49**
	Physicists in the fields…	53
	An equation difficult to conquer…	53
	From small-scale models to supercomputers	54
Chapter 8	**Maxwell's equations**	**57**
	Creative equations	61
	The Bible of electromagnetism	61
	A mischievous "Z^0"	63
Chapter 9	**The matter-energy equivalence**	**65**
	Between respect and awe	67
	Strange relations between space and time	67
	To believe, truly…	69
	The energy of the sun…	70
	… and of the atomic bomb	71
Chapter 10	**Schrödinger's equation**	**73**
	"Psy" or "Psi"?	77
	The first quantum vertigo	78
	Such a counter-intuitive world	79
	A wealth of applications	79
Chapter 11	**Heisenberg's uncertainty relations**	**81**
	At the heart of the quantum fuzziness	83
	An imperfect world…	85
	But our everyday world remains sharp!	86

A perfectly pure randomness 86
An operational "mechanics" but a mysterious one 87
Progress, an obsolete concept? 88

Chapter 12 Einstein's equations, general relativity 91

Dominating effects for cosmology 93
Describing the whole universe 95
Expansion and big bang 96
The evolution of the universe 97
Black holes: a laboratory... in thinking 97
Here and now 98

Chapter 13 Dirac's equation 101

And Dirac created antimatter 103
The equation first just fulfills its role 105
Quite physical solutions 106
Antimatter exists! 107
Believing in an equation 108

Chapter 14 Feynman diagrams 109

Tedious calculations... 113
A little diagram for a long calculation 113
A powerful language 115
Dreaming of Feynman diagrams 116

Chapter 15 The standard model 119

In the CERN cafeteria 121
The Lagrangian returns 124
Hunting the Higgs 125
A bright confirmation 126

Epilogue The limits of the standard model... and future physics 129

Standard, really? 131
But where is gravitation? 134
The ultimate equation? 135

Index 137

Chapter I

The law of light reflection

$$\theta_r = -\theta_i$$

This formula describes how a light ray behaves when it impinges on a reflecting surface, like the surface of a lake, or a mirror. It was proposed by Euclid, 3rd century AD.

This law is so simple that one could easily enunciate it without the help of an equation: it just says that to find the ray reflected by a mirror, one only needs to draw the line symmetric to the incoming ray with respect to the perpendicular to the mirror. However, deriving an "explanation" of what makes a reflection, is all but a fundamental evolution! Before the law, I "saw" an object and its reflection by the water of the lake, without asking myself any questions. Upon knowing the law, one part of what I see (the reflection) is "explained", linked rigorously to another part (the reflected object). To understand this relation, it is even necessary to question the very notion of "seeing"!

One can see light reflections everywhere. In nature, on the surface of water or on polished stones; in town on windows, on metallic surfaces, etc. Almost all of what we see includes reflections here and there, so much so that eliminating them requires some sophisticated technology: for instance, the anti-reflective coating of screens, glasses, or camera lenses.

But what is a reflection anyway?

We never think about the laws of light reflection. We never ask ourselves: but *why* does this reflection appear? We only start to think about it if we draw or paint, or if we want to generate an image with a computer.

In a natural environment, like a landscape in daylight, the unique source of light is the sun. I see a mountain or a tree because certain light rays emitted

$$\theta_r = -\theta_i$$

by the sun have hit this object and have been re-emitted towards my eye. Some light rays from the sun are absorbed, others are scattered in all directions: these complex processes are the origin of colors, of darker of lighter shades — these would only become understood much later.

But the case of reflection can be described in simple words: a light ray which hits a reflecting surface leaves it in a way symmetric to the incoming trajectory, exactly like a ball thrown against a wall. If we throw it perpendicularly to the wall, it bounces back to our hand. The more we throw it with a slant with respect to the wall, the more it will bounce away from us. This is what is said by the equation at the top of this chapter: the letters θ_i and θ_r denote the angles of the incoming (*i* for *incident*) and outgoing (*r* for *reflected*) light rays.

Once this law has been enunciated, I can "explain" why I see the reflection of a mountain in a lake: the mountain is lit by the sun, and it re-emits light rays in all directions. Some of these rays arrive directly into my eye: I see the mountain. But other rays arrive to my eye after having been reflected by the surface of the lake: I see its reflection. The above equation, and a bit of geometry, shows that the reflected light rays appear to me as an image of the mountain symmetric to that of the direct image with respect to the lake surface.

This looks perfectly familiar, and all painters can represent reflections, by intuition and observation. Explaining how the reflections form is not a revolution! Of course, this is only a partial "explanation": it only describes reflections, but not colors, diffusion, nor refraction… Neither does it say why some surfaces reflect and others do not. However, this small step is fundamental: never think of what we see as "devoid of interest" or "unexplainable", question even the most familiar words, like "to see" or "to watch". Attempt to interpret, to find links through simple laws, even at the cost of a deep upheaval.

By the way, what does "to see" mean?

I was very lucky with my physics teachers. In high-school, 11th grade, the teacher was a physics lover, who made a real point of performing delicate experiments: not in laboratory hours, but in front of the whole class, like a live show. For experiments with optics, we would switch off the lights, he lit up an arc lamp, crackling and stubborn (at the time there were no lasers, which would have made his life much easier…), then he introduced lenses or slits, he spun rotating mirrors, and the light rays streaked onto the classroom walls with strange ephemeral figures which he encouraged us to decipher. His contrivances almost always worked; great art!

But before these spectacular demonstrations, this teacher had spent half a class explaining that "seeing" is not to actively "have a look", but to receive light rays[1]. One full hour was dedicated to dismantle this intuition, so natural and so strong, which lies in the verb "to watch". It is not the look that reaches the object, but the light emitted by the object which reaches my eye. And by the way, one *cannot* "feel the look from somebody". I can only become aware that a person is looking at me when I see her eyes pointed in my direction. But if I do not see her eyes, her look can be directed elsewhere or "on me", I will never know.

Nothing sounds simpler, but I remember it as one of the first revelations of scientific thinking. Indeed, there are two ways of seeing the world. In the first one, I "watch", I "see" objects and reflections, without interpreting them in any way. In the other, a light source, a lamp, the sun, whatever, emit light rays. These are reflected, refracted, scattered by objects, then they arrive into my eye and then, only then, I see them.

Physics of enlightenment

So, this will be "doing Physics": leaving naive intuition aside, stepping outside the frame of subjectivity, to re-interpret the world. And preferably, doing it so such that this interpretation be common to all. Leaving aside "me", "I", one can enunciate laws, formulae. One can calculate and predict. All light sources emit rays, which reflect, refract, scatter on all objects, the same way for everybody. The laws of reflection and refraction announce the Age of "Enlightenment".

---

[1] Euclid formulated the correct law, but he was mistaken about the direction of propagation, and he thought that "light" was the action of the eye!

Chapter 2

Snell-Descartes law... and the Principle of Least Action

$$n_2 \sin \theta_r = n_1 \sin \theta_i$$

After the reflection of light, here comes the law of refraction, which describes what happens to a light ray which crosses from one transparent medium to a different one, when it is not reflected. Here again, examples are plentiful, and most often we do not pay attention to them. Perhaps we do find a bit strange that a pole stuck in a pond appears broken at the level of the surface, at the interface between air and water…

In the domain of light, direct applications of this equation are innumerable, since it reigns over all optical lenses. But we will see in this chapter that this simple law, discovered in a specific case, would later be the source of a new, very general and very powerful way of comprehending Physics: the "Principle of Least Action".

The equation indicates that the angle of the ray leaving the interface is different from that of the incoming angle, which gives the illusion of a broken stick. You probably know the mathematical expression "sin": it is the sine function which one can compute with a button of any scientific pocket calculator. Historically, it took some time to understand that this was the correct function, and not simply the value of the angle itself for example, but once the law is known, the equation is quite simple and practical.

Experimenting

Equipped with this equation, we can carry out the refraction experiment using various pairs of transparent media: air/water, air/oil, water/oil, etc. We find that the numbers n_1, n_2, n_3 are each characteristic of a medium: n_1 for air, n_2 for water, n_3 for oil. This is the refraction index of the medium. Once we have measured these numbers, we will be able to predict the refraction at

$n_2 \sin\theta_r = n_1 \sin\theta_i$

the interface for any pair of media, and to use this prediction to build any optical system we want.

Practical applications are numerous, first to understand nature, like rainbows or how the eye functions, but most importantly to create optical instruments, from a magnifying glass to a microscope, not forgetting eyeglasses; they are all designed based on this equation.

Whose law is it anyway?

The story could have stopped there, or rather it could have remained limited to the evolutions of optics: microscopes, astronomical telescopes, photo lenses, etc.: all great practical successes! But there is more, an even more interesting consequence, because from this equation would rise a new development, more abstract and very important for theoretical physics.

The law was clearly enunciated around 1620–1630, independently by both Snell Van Royen and René Descartes. Who was first is only one of the many debates between the English and the French, a useless debate since the construction of the law spreads over several centuries. It began as early as the 10th century in the Arab world, then several great minds contributed their share, including Johannes Kepler.

When light loses no time

The development I was talking about above was found by Pierre de Fermat around 1660. He discovered that he could retrieve "Descartes' law" (he did not know about Snell…) if he started from a very simple principle: assuming that the velocity of light depends on the medium in which it propagates, the light ray takes the path that enables it to spend the least possible time between two points located on each side of the interface. The quantities n_1 and n_2 are then interpreted as the inverse of the velocity of light for each of the media.

You should imagine that in the middle of the 17th century, the value of the velocity of light was still unknown and some even maintain that it propagated instantaneously[1]. Of course, Fermat had absolutely no way of measuring this speed in different media: to travel one meter in vacuum, light only takes 3 billionths of a second, and 1.4 times more in water, i.e. 4.2 billionths of a second. These durations were much too small to measure in his time. Nevertheless, Fermat not only discovered the "principle of least duration", but he showed that the principle works if the velocity of light is smaller in water or glass, than in air, by a factor of 1.2 to 1.5. Ironically, he thus opposed Descartes, who thought that the speed of light is higher if the medium is denser. But the principle worked beautifully, confirming Fermat's hypothesis.

The intellectual progression which guided Fermat towards his principle was originally an intuition, or clear-sightedness: "Nature always acts along the shortest and easiest ways". This idea, more precisely this way of formulating a physical law, would turn out to be extremely fruitful: instead of writing an equation which relates two angles — a mechanistic equation — one enunciates a general principle which states that a light ray chooses the path of minimal duration.

A principle of "least action"

The method was adapted to mechanics by Pierre Louis Moreau de Maupertuis in the 18th century, namely to the movement of objects submitted to forces, and was named the "Principle of Least Action". It would be further formalized by Leonhard Euler and mainly by Joseph Louis Lagrange, still applying it to mechanics. In this framework, the notion of "action" was introduced: it is the product of an energy by a time. The path of "Least Action" is the path which takes the least time, or which modifies the energy by the smallest amount, or the best combination of these two: the most economical path, the "easiest" as Fermat said. Regarding Fermat's optics, the energy of the light ray is constant, hence the economy is made only on the time. For Lagrange's mechanics, the "path" is understood in a general sense,

---

[1] Not for very long: the very first estimate of the velocity of light was proposed in 1675 by Rømer and Huygens who observed the satellites of Jupiter, another incredible inference. A very imprecise value (220 000 km/s), but not infinite, and of the correct order of magnitude!

since this encompasses the evolution of the whole system: the position and velocity of the objects, the energies stored here and there (for example the tension of a spring). To describe the evolution of a mechanical system, even a complex one, one can just assert that it should obey the "Principle of Least Action".

Then, in the 20th century, scientists realized that the new physics, Quantum Mechanics, which we will talk about later, could be beautifully formulated by a principle of least action. In the quantum world, everything is somewhat fluctuating. The positions and velocities of particles are often imprecise, and are not the best quantities to use for the calculations. More global quantities, like the different forms of energy, are clearer and easier to exploit.

If one can properly define the various forms of energy, the Principle of Least Action allows to predict how they will transform into each other during the interaction of particles, or how they balance each other in an atom for instance: by "the easiest ways" as Fermat quoted, or in modern words by "following an evolution which tends to minimize the action at all times".

A formula to sum everything up?

Indeed, so much so that this formulation now dominates particle physics: all theories, from the Standard Model (chapter 15) to the most exotic theories, are expressed in this framework. Each theory defines its objects, its particles, its fields, its forces, its energies, and gathers these definitions in a big formula called the "Lagrangian", in honor of the work by Lagrange that we saw above (we will see a few examples of Particle Physics Lagrangians at the end of this book). The design of the theory can stop here. Do you want to apply it to a concrete case? You need "only" express the Principle of Least Action, then run through the calculations and there you have it!

Beginner physicists learn the modern form of the principle, formulated by Euler and Lagrange. This is a beautiful equation:

$$\frac{d}{dt}\left(\frac{\partial L}{\partial \dot{q}_i}\right) - \frac{dL}{dq_i} = 0$$

The letter "L" represents the famous "Lagrangian", the key physical quantity from which everything can be derived. The "q_i" and "$\dot{q}_i$" are the positions and the velocities of the objects respectively. As you can see, I have not included it as a fundamental equation of this book, because it is not truly an "equation of Physics". Rather it is a procedure, an algorithm, and not a vision of the world.

I remember when I was a student in Physics, we were all excited when we studied the principle of least action, the Euler-Lagrange equation, and its quantum equivalents. The formula looked magical: you wrote down a nice Lagrangian, you turned the crank and there it was, you obtained the solution of a complex mechanical movement or of the interaction among various particles. But quickly came the disenchantment: the equation does not say what the objects are, give their energies nor their interactions. Only if you have defined your vision of the world, and if you managed to summarize it in the form of a Lagrangian, then the equation will tell you how your theory will apply to concrete cases.

Hence there is no miracle, but the formulation "Lagrangian + Principle of Least Action" strongly influences physicists. And maybe also everyday life? Isn't it exactly what psychologists presently tell us? To find happiness in an economical lifestyle? To work efficiently, to avoid useless conflicts, and to not remain enslaved to the dispensable? If only we knew the Lagrangian of Happiness…

Chapter 3

The fundamental principle of dynamics (Newton's second law)

$$F = ma$$

What does this very simple equation say? It says that the force is the product of the mass by the acceleration. I am aware that the last sentence does not mean much to the non-scientific reader. It might become much clearer if I write it in the equivalent form:

$$a = F/m$$

and if I say in plain words: "When I apply a force F to an object with mass m, I provide it with an acceleration a, directed along the direction of the force and with a value of F divided by m. In particular, if I apply a force with twice the intensity, the acceleration will be doubled. If I apply the same force to an object with twice the mass, its acceleration will be twice less." This is quite simple indeed.

Or is it? Historically, it took centuries to write this simple formula. The long and difficult part was not to invent the form of the relation between the variables:

Something = this "times" that

What was really important was to reach a clear definition of the relevant concepts to describe the mechanics of bodies by a simple relation. This definition is by no means trivial…

Perfect because motionless (?)

What defines a movement? It is natural to think of speed, or of the changes in direction. But how can one give a precise meaning to these intuitions?

F = ma
12

It is even more subtle to describe the mechanical action which I apply to an object. How can I quantify this action? What is a force, precisely?

Can it be defined in a universal way? Is there anything in common between my hand that throws a stone and a cannon which fires a ball?

Before Galileo and Newton, the motion of a body was subject to a vague, ill defined, language, dominated by the influence of Aristotle. For Aristotle, perfection was akin to immobility. An object on which nobody acts, is motionless. With $F = ma$, if there is no force the acceleration is null, meaning that the speed does not change. In other words, an object which originally moves with a velocity v and on which nobody acts, continues to move in a straight line at a constant velocity.

From fountains to planets

This is a true revolution. Think about planets: don't they move on closed trajectories, which have them periodically come back to the same locations in the sky? They may rotate around the Earth — before Nicholas Copernicus — or around the sun — after Copernicus, in either case they do not follow a straight line in space, otherwise they would never come back and visit us again. Hence something, some "force", acts on the planets. In a way, they are no longer "ideal" along Aristotle, free from all influence. As planets are associated to divinities — Venus, Mars, Mercury, etc., by writing $F = ma$ mankind makes an incredibly audacious move: they take away some brightness from the gods…

I once saw a very striking image: A Renaissance draftsman had represented the trajectory of a cannonball, and decorative fountains, in the same drawing. The cannonball rose along a slanted straight line, then it suddenly fell vertically. But the water jets from the fountains followed nice smooth curves — parabolas, as would be understood much later. The draftsman was not aware that the drops in the water jets and the cannonball should have followed the same type of trajectory. He drew the water jets as he saw them, and the cannonball trajectory, too fast to be visible, as he *thought* it was. Of course the real trajectory of the cannonball is indeed a parabola, like the water jets.

But why did he draw this strange triangular trajectory? The usual explanation at that time was the following: at the start, the cannon communicates to the ball a certain amount of "impetus", then this impetus gets used up, and when none remains, the up going straight-line stops and the ball falls vertically. But what is this initial impetus? How was is transmitted to the ball? Why and how is it spent? When will it be used up? To correctly describe the trajectory of water jets, cannonballs, and planets, one will need to understand forces, mass, an acceleration. Fortunately, once these quantities are well defined, the equation which relates them is very simple, as we can see.

A touch of nostalgia for the lost "γ"

In the equation above, "*a*" is the acceleration, that is the variation of the velocity. Every driver knows that to accelerate means to increase one's speed. Braking is also an acceleration, but negative. The velocity, or speed, measures the variation of the position with respect to time: at high speed, the position changes more for the same duration. Thus, the acceleration is the variation of the variation of the position, with respect to time.

This "calculus of variations", as it was called in the 18th century, is at the basis of almost all Physics. If I change "this" by a small amount, then let us see by which other small amount "that" changes. This is quite a modest attitude indeed. I do not try to explain directly how things are, I only attempt to describe the relations between variations of positions, velocities, energies, temperatures, etc. I say nothing (yet) about how the planets were created, but I am happy to understand their motion, and most importantly, to be able to *predict* it. I observe their positions today, I understand the variations, hence I can compute the positions they will have tomorrow, in a year, in a century, in an eternity.

As a high school student in Paris, I first learnt this equation in the form $F = m\gamma$. I am not sure why traditionally in French high schools the Greek letter "γ" was used[1]. It has since been replaced by the simple "*a*" for "acceleration", more like the international notation, and less impressive for the students.

But I liked the "γ" a lot. It was one of the very first symbols found in a physics class in high school. To me, it expressed that it is possible to

[1] Maybe "γ"was introduced to generalize the notion of acceleration, from the best known one: "g" for "gravity".

represent a physical quantity, a concept, in a clear and powerful way. The symbol of symbols, so to say! Anyway, it is one of the very first symbols I began handling when learning physics, then later in my work as a physicist. It was also the mark of a kind of awe: perhaps it was not so simple; perhaps a "γ" deserved a bit more reflection than an "a"?

A rather dynamic physics

And there was more: with $F = m\gamma$, there came the first *differential equation*. A relation between variations... Here was the real physics, not grocery accounting anymore — my apologies to grocery keepers. In the previous classes, mechanics was only used to calculate the equilibrium of a scale, the pressure at the bottom of a swimming pool. This was "statics", cold, rigid, boring, and accompanied by baroque explanations. How many students had given up physics after being explained that *a table stands upright because of the reaction exerted by the ground on its feet*? What? Which twisted mind conceived this idea that the ground pushes on the table from bottom up? Was it really necessary to define the concept of force to understand that?

$F = m\gamma$ was dynamics, and for all bodies. The movement of the air, of vibrations (at that time *good vibrations*), acoustics, audiocassettes... The political movement, the Great Leap Forward, the hope for a better world... Movement, everywhere. Everywhere, slender differential equations gave the world an inexhaustible vitality.

The prize books of my youth, full of fast trains, flying automobiles and space rockets, were still there, despite the doubts which started to rise in me: questions about consumer society, emerging ecology...

But who cared? If it was not going to be this society it would be another one, just as animated, submitted to competition and with rapidly varying forces, *dynamic* in any case, like one is at the age of 16 or 18. Like one could be at 18 in these years without crisis, without unemployment, without housing difficulties. Like the youth of today could be, if we only gave them the possibility.

Eventually, too, immobility, or rather the uniform motion in straight line, seen not as an absence, a hollow, a disenchanted inaction, but like an equilibrium of all the forces, peace amid turmoil: the first appearances of Buddhism and Zen in our society, the peace and love movement. Statics understood at last, as the ultimate state of dynamics.

Chapter 4

The law of gravitation (Newton's first law)

$$F_G = -G\frac{mm'}{d^2}$$

This is the law of universal gravitation, the force which acts on heavenly bodies and which "made the apple fall on the head" of Newton, its inventor in 1684. It claims that all bodies mutually attract by gravitation, and it gives the magnitude of this attractive force as a function of the distance between the two bodies.

Hence this law is a monument, since it was the first to relate the movement of celestial bodies and gravity on Earth, two phenomena which à-priori have nothing in common. It would then allow to understand, and predict, the movement of all celestial bodies.

What does the equation say? It states that two bodies of masses m and m' mutually attract, and the force that each exerts on the other is proportional to the masses m and m' of the bodies, which sounds logical, and inversely proportional to the square of their distance. In simple terms, this means that if we pull the bodies apart and multiply their distance by two, the attractive force between them will be divided by four. The power of this law resides in this simple dependence, which has since been perfectly verified in Nature (at least until recent times): the motion of the moon, of planets, comets etc. can be calculated with excellent accuracy.

The unloved equation

Strangely enough, I never really liked this equation. Scientifically, I cannot deny the huge progress it represented. I certainly admire the formidable

$F_G = -G\frac{mm'}{d^2}$
1320
3248
G
SUPER BONUS

power of the mind which formulated it. But this law of universal attraction paradoxically never attracted me, and I am not sure exactly why.

If I were presumptuous, I would say that from the beginning I had the intuition that such an action at a distance sounds a bit suspect. According to the law, for example, the Sun exerts a force on the Earth. The magnitude of the force is almost constant with time, since the Earth orbit is almost a circle, and its distance to the Sun is almost constant. But, according to this law, if I removed the Sun, its influence on the Earth would vanish instantaneously. And this is suspicious. How can such an action be immediate, at such a distance?

But I should remain modest: clearly, I am reimagining my feelings of that time, in light of all I have learnt since. Specifically, I have since learnt Einstein's relativity, which forbids instantaneous action at a distance: no particle, force, or information, can propagate faster than light.

Too simple a law?

Another aspect which I find a tad disappointing in this law, is the dependence of the force F with the distance d between the bodies, namely like the inverse of its square. Naively, this dependence with $1/d^2$ may appear remarkable, as if it unveiled a beautiful and fundamental truth. But if you think more, it is only the most trivial form for a force exerted by an object on other distant objects.

Here is why: looking from far enough, the source object appears point-like. Let us imagine a sphere of radius d, centered on this object. The area of the sphere is $S = 4\pi d^2$. Thus if we increase the sphere radius by a factor of two, its area increases by four. This is exactly the inverse of the dependence of the force on the distance. The fact that the attractive force exerted by an object at a distance d decreases like $1/d^2$ simply means that the influence of the object gets uniformly diluted in space. At a larger distance the influence is weaker, quite intuitively. It is weaker because *there are more points to reach* — like the number of points on the surface of the sphere. In short, a dependence of the form $1/d^2$ is the most trivial form one could imagine for the force exerted by a point-like object. Indeed, the same law describes the force between two electrical charges:

$$F = k\frac{qq'}{d^2}$$

A law discovered much later by Charles Augustin de Coulomb (1785).

… to conceal ignorance?

Despite their precision and their immense success, Newton's and Coulomb's laws first express… an ignorance! I don't know what a mass or an electric charge is. I don't have the slightest idea of their origin nor of their inner mechanism. But I can gather all the influence of an object's gravitation in one quantity: its mass. The form of this influence is then the most simple, as a function of the distance. And, similarly, for the electric charge and electrical forces.

Later, General Relativity would provide gravitation with a more ambitious program. There, mass, energy, and the shape of space-time are linked in a consistent system. Einstein's theory went beyond the "easy" principle of lumping everything in the single property of mass, without asking more questions. From this coherence, would stem new unexpected objects, like black holes, as we will see later.

E pur, si muove!

Since I learnt all this long after having discovered Newton's equation, it is vain to pretend that I never liked it because of precocious intuitions. Then why? Perhaps, simply because when it is taught in high school, this equation is just useless. Students are simply unable to use it! I mean to *really* use it, for the application for which it was first invented: to predict the motion of planets, like our good old Earth.

The problem looks simple. Two tools are now available: on one hand $F = ma$, on the other hand the inverse square law for the force. We only need to say that the gravitational force exerted by the Sun on the Earth,

$$F = -\frac{(Gm_S m_T)}{d^2}$$

is the force which induces the movement of the planet through $F = m\gamma$. Then the movement of Earth obeys the equation:

$$m_T \gamma = -\frac{(Gm_S m_T)}{d^2}$$

Where γ is the acceleration of the Earth (including its direction) and d is the distance of the Earth to the Sun. Solving this equation leads to the understanding of the motion of the Earth, the shape of its orbit, its velocity at each instant, the time it spends close or away from the Sun.

Fine, but this resolution is not so easy and it requires mathematical tools more advanced than those which are taught in high school. Therefore, in high school, the application is limited to exercises without much practical interest, like calculating the force between two fixed objects. With elementary mathematics, it is possible to solve the equation with the approximation that planets run along circular orbits around the Sun. But is well-known that this is not the case (the orbits are ellipses), and moreover, with this approximation, all the power of the law is lost. We should better assume that the Earth is bound to the Sun by a piece of string, then we will obtain the same circular motion and everything will be even simpler. Only later, in university, did I learn the proper resolution of this equation of motion, understanding the elliptic trajectory of planets and comets. But this did not really improve my feelings towards Newton's law.

Cosmic chaos

The motion of a single planet around the Sun, as described by this law, is for me a monument, but a charmless one. In contrast, the study of the movement of all the planets of the solar system reveals unexpected and fruitful questions. For one planet (say Earth), the main influence is the attraction of the Sun, which induces an elliptical, but also periodic movement: under this single influence, the orbit would never change, and every year the Earth would come back exactly to the same point — provided the characteristics of the Sun did not change. This would be the case for each planet, which would travel along its own ellipse for eternity.

But the planets also exert some influence on each other. As the distances between planets vary rapidly, these mutual attractions induce very small but very fluctuating perturbations. The question is if these perturbations, accumulating over time, could significantly modify the primary motions along the ellipses. If yes, after which time interval? And is the perturbed movement still periodic? Or at least, sufficiently periodic that it can be predicted in the long run?

The question is so delicate that its answer has varied several times since Lagrange formulated it in 1766. Without going too much into details, nor following its historical evolution, today scientists think that the Solar System is "chaotic", in particular thanks to the work of Jacques Laskar (1989). It is impossible to predict the orbits of planets, and even less the position of planets along their orbit, beyond a few tens of millions years. And this although the only forces at play are Newton forces between the Sun and the planets, and those among the planets, and these forces are perfectly continuous and without any randomness.

When we solve the equations of motion and we project 100 million years in the future, the solutions become infinitely sensitive to the initial conditions, namely the starting positions measured today. And even worse, this sensitivity is not progressive, but it appears as an impenetrable wall. According to Laskar, an uncertainty of 15 m on Earth's initial position induces an uncertainty of 150 meters on the predicted position after 10 million years, which could be deemed as acceptable, but an uncertainty of 150 million kilometers after 100 million years! In other words, we cannot predict the position of the Earth on its orbit at a chosen date. This ultra-fast increase of the uncertainty with time makes illusory the idea that to predict the position at a chosen time, it is only necessary to measure it now with enough accuracy.

A planetary pool game

One can even ask the question of the global stability of the Solar System: since it is impossible to predict the long-term position of each planet, can we at least know if the orbits will be roughly similar? The present orbits are elliptical, but rather close to circles (their "eccentricity" is small, in physicists' jargon), and considering the differences between their average radii, it is clear that they do not intersect. But what about the long run? Would it be possible for the orbits to deform sufficiently to allow for intersections, and, subsequently, collisions? Could it happen that a planet would pass so close to another one, that one of them would be ejected from the Solar System? In short, could our usual peaceful roundabout transform into a scary planetary pool game with impacts and bounces?

Recent estimates, for several billion years in the future, carried out with numerical computations on supercomputers, show that this possibility does exist, although it is rather improbable — of the order of 1%.

As I said above, the answer to the question of the stability of the Solar System has already changed several times, between "stable" and "non-stable". The chaos described by Laskar is the answer by today's science. Could there be other regulating phenomena, not yet accounted for? These are indeed lively and fruitful questions — this type of chaos is also present in many other physical systems — which are much more exciting than the resolution of Newton's equation for a lonely planet…

Chapter 5
The ideal gas law

$$PV = nRT$$

This is again a very simple equation, which describes the behavior of many common gases, including ambient air. This law is in itself very important, since in nature it allows to understand the variation of atmospheric pressure with altitude. In the laboratory and in industry, it opens the way to a quantitative handling of gases: air, vapor, etc.

But beyond these direct applications, it will play an unexpected role. Although it is meant to describe the behavior of a human-size volume of gas, or larger, the nature of this equation will be decisive in claiming that matter is made of minuscule and almost independent atoms.

An ideal air

I will emphasize this little-known aspect later in this chapter, but let us come back to the equation. It says that for a fixed quantity of gas enclosed in a box, the product of the pressure P by the volume V is proportional to the number of atoms n and to the temperature T. The coefficient R is a proportionality factor, called the "ideal gas constant".

In simple words: if I reduce the volume of a given quantity of gas (slowly, so that the temperature does not change), its pressure increases (this is indeed what happens in a bicycle pump). If, in a fixed volume, I inject more gas (increasing n), the pressure increases (injecting air into an already inflated tire). If in a given volume, I increase the gas temperature, its pressure increases (think of a pressure cooker), and so on.

$PV = nRT$

Good fortune

The equation is not sexy at all: no variations, no pretty indices or exponents. But it is incredibly efficient. By chance in fact… Because most usual gases, at ambient temperature, are almost ideal! Fortunately for us, and for our desire to understand the world around us, nature offers us most of the bodies in three forms (the technical word is "phases"): solid, liquid, and gaseous. Not under six or twelve forms… And regarding the equation which relates pressure and volume (at constant temperature), nature could not have made it simpler: in a first approximation, solids and liquids are *incompressible*: whatever pressure I exert on them, their volume does not change. Then there are the gases, for which $PV = constant$: if I decrease the volume by a factor of two, the pressure increases by a factor of two. Really, really simple.

Today, we know the conditions under which gases are not ideal, for instance when they are strongly compressed, to the point where they almost become liquids. However, these conditions are not very frequent in the world around us. This apparent simplicity has thus been essential in allowing us to understand this sector of Nature. If instead of three phases, there would have been twelve, each governed by an *equation of state* more complicated than $PV = nRT$, could we have developed the physics of matter and thermodynamics? Could we have gone directly to complex descriptions if Nature would have imposed it, without first providing a stage with a small number of simple cases?

The air of the law?

What is it that enables us to state physical laws? Let me propose three answers, all quite different in spirit:

— nature possesses real regularities, fundamental simplicities, which we discover and then enunciate. This is good old realism.

— Whatever the complexity of nature could be, we would be able to point to some regularities, to make some approximations. Our physical laws would not truly reflect the true nature of our world, but only those regularities which we would be able to formulate. This is a form of philosophical relativism.

— A thinking being, specifically one able to formulate physical laws, can only exist if nature contains certain regularities. An excess in complexity at all scales of space and time would only induce chaos, and would not allow evolution, nor learning, nor the transmission of knowledge from generation to generation. Hence our own existence means that there should be constants, regularities, laws. This is a form of the "anthropic principle".

These epistemological debates are endless and are not the main topic of this book. But think about this very simple example, $PV = nRT$... Which one comes first? The ambient air, this (almost) ideal gas, or my definition of pressure and temperature?

An amazing hypothesis

In any case, the law of ideal gases opens a vast field of explanation for everyday phenomena: the barometer, the variation of atmospheric pressure with altitude, etc. Furthermore, this Law has numerous practical applications: the hot air balloon, the steam engine, pumps, chemistry and so on. The Ideal Gas Law also had a large impact on the fundamental understanding of the physical world. In the 18th century, the debate between "continuous matter" and "atomistic matter" was not yet settled. Ever since the classical Greek era, around 400 BC, Democritus claimed that matter is made of atoms, tiny invisible elementary bricks. Before him, Leucippe had made a similar claim, and in India a certain Kanada as early as 600 BC.

Then the atom hypothesis lasted for centuries, although absolutely no sign of it had been evidenced. Even the invention of the microscope, at the end of the 16th century, could not allow to "see" the atoms. For a good reason: we know today that visible light has a wavelength much larger than the size of an atom. By the way, this is why our eye can see it, once again the old chicken-and-egg problem!

Back to our atoms. Although it was never demonstrated before the 19th century, the atom hypothesis always had adepts, and among them, the best minds. Newton was an atomist. Even Shakespeare knows the atoms:

"It is as easy to count atomies as to resolve the propositions of a lover[1]", which proves that he estimated their number in matter to a very, very large value…

The atom hypothesis survived in parallel to other explanations of the elementary constituents of the world, like Earth, Wind, Water, and Fire. At the beginning of the 19th century, John Dalton gave a clear formulation, by only considering the chemistry of elements, especially the proportions with which they assemble or separate during chemical reactions. However, for Dalton, this was only a handy model, without any experimental proof of the existence of real atoms, and of course without any indication of their possible size and number in matter.

Measuring the size of atoms by blowing air

A decisive step was made, around 1860, when James Clerk Maxwell and Ludwig Boltzmann proved that a gas made of very small atoms (or molecules), with few interactions among them, and only agitated by temperature, would behave precisely like an ideal gas: $PV = nRT$.

At this stage, the model did not say anything about the size of the molecules, and in turn nothing on their number in a given quantity of gas (what is called today the "Avogadro number"). But the trend was launched and the reasoning went on. If it is made of atoms, matter is a gas when these atoms are rarefied: far from each other. When the gas is compressed, the atoms get closer and when they "touch", condensation happens, into a liquid or a solid.

Hence atoms would have a size! But how could one estimate it? By measuring the *deviation* from the Ideal Gas Law: if the atoms are point-like, they ignore each other, the gas is ideal, and consequently its viscosity is zero. If by contrast they have a finite size, the atoms will undergo collisions, and these collisions will induce a viscosity. In his atomist model, Maxwell achieved the calculation which could relate the viscosity of a gas to the

[1] In "As you like it". In the play "lover" is used for a woman. To be perfectly honest, *atomies* could also refer to dust particles suspended in air, that glitter in a sun ray. But the reference to atoms is allowed…

average distance between two collisions of molecules (what physicists call the "mean free path").

Josef Loschmidt, in 1865, measured the viscosity of air, by blowing air through a small hole and measuring the output velocity. Using Maxwell's relation, he was the first to deduce the approximate size of air molecules. This step, not often mentioned in textbooks, is still extremely important: for the first time, matter was considered as truly made of atoms, with a typical size of one 10 000th of a millimeter, and about 3.10^{22}(30 000 billions of billions!) molecules in one liter of ambient air. He reached this first estimate without any electronic microscope or anything alike, in fact without ever having access to an isolated atom! Only by comparing Democritus' intuition, more than two thousand years old, to the Ideal Gas Law.

A revolution in progress

I am still amazed that this atom hypothesis could rise, then survive for so long without the slightest proof, over more than twenty centuries. During most of this period, gases were not even identified as such. It was clear that a solid body transforms into a liquid when melting, but what happened when the same liquid body evaporates was not understood.

Probably, the fusion of solids into liquids was the best argument for the atom partisans. When we heat a solid body and it melts, transforming into a liquid, it does not lose its nature. It is clear that this is the same body under two very different forms. The simplest way to apprehend this transformation is perhaps to assume that the body is made of atoms, and that only their arrangement changes between solid and liquid.

In a sense, the atom hypothesis amounts to a Copernican revolution. At first glance, matter is continuous: it can be divided infinitely. God is also infinite and perfect, and this smooth and continuous perfection does not need any particular *elementary structures* or *size scale.* God is also infinitely complex. Nature, which He created, needs not be reducible to elementary bricks of a small number of types: it is in its complexity as He created it. Reductionism, according to which the apparent diversity of nature comes from the variable arrangement of a small number of elementary constituents, ultimately opposes a divine syncretism.

Chapter 6
Hooke's law

$$F = kX$$

Not famous, since it is rarely seen in school, Hooke's Law says that the lengthening of a metal piece is proportional to the force with which we draw on it. I have chosen it for its simplicity, and the fact that it represents the beginning of a movement towards rationalization, bringing mathematics into physics and engineering. Think about it: we model the behavior of a small piece of matter, and thanks to the power of differential algebra, it becomes possible to calculate how one should design and arrange struts to build the Eiffel Tower!

The equation is rather similar to that of the Ideal Gas Law we saw in the previous chapter, at least in its $PV = constant$ version, formulated by Boyle and Mariotte at the end of the 17th century: the pressure of a gas increases as much as its volume decreases. This similarity — a simple proportionality relation — is not there by chance: Robert Hooke and Robert Boyle were close collaborators.

The physics of the industrial revolution

I have to admit that the physics of continuous media (solids, liquids, and gases) is not my specialty. My vocation has always been to understand the infinitely small, specifically what the elementary bricks of matter are and the nature of their interactions. I have taken great pleasure in studying atoms, nuclei, electrons, neutrinos, quarks, and all their cousins. A great pleasure too, in becoming initiated to "modern" theories like Quantum Mechanics or Relativity, both invented at the beginning of the 20th century. It was such an emotion to discover that matter does not behave like our intuition suggests,

$F = kX$

with its particles which are everywhere at the same time, and a travelling twin who ages less quickly than his motionless brother!

But let us be fair: the development of technologies, the industrial revolution, and the corresponding drastic change in lifestyle, all result from the previous two centuries. The steam engine which provides mechanical energy, the calculation of metallic pieces to build bridges, buildings and ships, or the invention of the refrigerator and of the internal combustion engine, all this did not wait for Quantum Mechanics to arise. Neither did Chemistry, which I acknowledge as fully essential, but for which I feel rather incompetent...

Many books investigate the technical, economical and sociological origins of the industrial revolution. From a physicist's point of view, it is striking that this revolution was not due to a radical change in the understanding of matter at that time. Atoms were not yet discovered, matter was seen as smooth and continuous. It simply appeared under three distinct states: solid, liquid and gas, with water being the simplest example.

One could associate to matter some intuitive but ill-defined quantities: temperature (hot or cold), hardness (hard or soft), fluidity (fluid or viscous). From a physicist's perspective, all this revolution only consisted in defining these quantities better and, in parallel, applying to matter the new mathematics discovered by the work of Newton and Leibniz, namely the calculation of "variations", today called calculus. Thus, without asking "what is matter?", one has modelled its behavior better and better, first in specific cases, then in a more general way, by combining these various models.

Elastic like a metal

Again, Hooke's Law is a very naïve kind of model: one could not imagine a simpler dependence! Indeed, there are many bodies for which the law does not apply: wood, which gets torn apart if it is drawn upon, or stone which fragments is we exert too much pressure on it. But, like for the Ideal Gases, a "good fortune from nature" decided that the law works very well for most metals. We say that a metal is "elastic": if we pull on a piece of iron, its length increases proportionally to the applied force; if we stop pulling, it returns to its original length, like a spring.

It sounds so naive, but equipped with this very simple model, and applying calculus, one can predict how a metallic beam bends under a load and which forces propagate through a more complex structure. Nothing more is needed, almost, to build the Eiffel Tower!

The "mechanical power of fire"[1]

Calculus found a vast and fertile field of application with the development, in the 19th century, of a new science: thermodynamics. The starting point was the invention of the steam engine, which was able to provide mechanical work much beyond that produced by men, animals, or even wind and water mills. Furthermore, this work could be provided at a small or large scale, in a farm or in a steel factory, even in motion like aboard trains and ships.

In half a century, the use of heat to provide mechanical work became universal and polymorphous. A method was also invented to use mechanical work to remove heat: giving us the refrigerator! Then the direct use of gasoline in the internal combustion engine further increased the human craving for fuels, with the consequence today that all this burnt fuel is now in the atmosphere in the form of CO_2...

Heat, movement, energy

The employed fluids, like water vapor, air, then all kinds of gases and liquids, are natural compounds, each with their characteristic features like density or heat capacity (which amount of heat is needed to raise its temperature by one degree). To use them optimally, with good efficiency, or to design new systems, it was not necessary to predict these features from universal principles or microscopic laws. But it was essential to model them well, that is to identify their intrinsic properties, which do not depend on the considered system, then to measure these properties with a good accuracy. In those days,

[1] "Réflexions sur la puissance motrice du feu", a book by Sadi Carnot, one of the founders of thermodynamics.

nobody knew *why* water appeared under three states (ice, liquid, vapor), but one precisely measured for which conditions of pressure and temperature water changed from one state to the other, or which amount of heat was needed to melt one kg of ice.

It was then possible to link the behavior of complex systems, like a steam engine, a refrigerator, or a compressor, to quantities measured in the laboratory on samples of matter. The intuitive quantities became better and better defined, like temperature and heat. New ones appeared, much less intuitive, like the mysterious entropy, proposed in the second half of the 19th century.

The theory which accompanied these developments was thermodynamics. Seen by a physicist "of the elementary", it is not really a theory: it does not try to explain the world, but rather to exploit it. The mathematics involved were essentially generalizations of differential calculus: differential equations, most frequently linear ones, meaning that the variation of one quantity is related to the variation of another quantity by a simple coefficient. For instance, let us look at a small amount of water vapor, and measure how its pressure varies if we heat it or if we change its volume. We need to be careful with the definitions here: do we heat and let the volume inflate, or maintain a constant volume? Do we change the volume and let the temperature grow or to we keep the temperature fixed with an external heat bath? And beyond variations: what about more complex systems?

Again, thermodynamics is not my strongest point, and throughout my education I was a bit suspicious of these long series of equations which manipulate pressure, volume, temperature, energy and entropy in turn… I had to learn a pedantic usage and some applications, but I seldom saw the meaning of it. Why is this quantity fixed? Why is that one variable?

Today, I can find on the Internet the thermodynamics course of a colleague and friend[2], which contains the following lines:

"This annex, which is extremely theoretical, shows how to do thermodynamics, without understanding (almost) any of the underlying physics".

…which made me smile sincerely, remembering my first contacts with "Thermo", as we used to nickname it.

---

[2] Patrick Puzo; thanks for the jest!

Fortunately for me, in the next classes we learnt about Statistical Physics, which explained all this as the collective behavior of a large number of atoms (or molecules), and the concepts were then clarified. Nevertheless, this analytical, laborious, thermodynamics still allows for designing car engines, fridges and air conditioners.

Looking at the Boyle-Mariotte Law or Hooke's Law, I am again fascinated by this intellectual construction and the simplicity of their formulation. Can everything in nature be described by simple proportionality relations? In fact, there exists a whole branch of science which deals with "complex" phenomena, where the relations between variables are not proportional. They are called non-linear, and strongly so: a small variation in one quantity drives an enormous variation in another.

In any case, all our human (and scientist) intuition, is based on these simple kinds of relations: I pull twice as hard, the spring lengthens twice more; I go twice faster, it will take me half the time; I buy twice more things, it will cost me twice the money — although in this last case, everybody has a good notion of a "non-linearity", since if I buy twice more things, I can hope to spend *less* than twice the money!

Which constraints do these simple relations impose on our way of seeing the world? Can we (really) think about complexity? For sure, the media does not help in freeing oneself from simple minded thinking: they like to represent the whole economy, and even life in general, with percentages. From the rate of unemployment to romantic encounters, not forgetting the consumption of sodas or books, do percentages truly reflect reality?

Chapter 7
Navier-Stokes equation

$$\frac{\partial u}{\partial t}+(u\cdot\nabla)u$$

$$=-\frac{\nabla P}{\rho}+\mu\nabla\times(\nabla\times u)+F$$

From an aesthetic point of view, this is without doubt a "beautiful equation". You can appreciate the "high level": a few Greek letters, a couple of very elegant "∂", and several occurrences of the strange symbol ∇ — "nabla" for the enlightened. The equation is written here in its modern notation proposed around the end of the 19th century, much later than its invention which dates to the beginning of that century. The modern notation ensures a compact and elegant form, for a set of equations which would look repetitive without it.

A quite scholarly look indeed, but in substance, it is still a very "descriptive" equation. Basically, it amounts to adapting $F = ma$ for a fluid. Newton's mechanics describe the motion of rigid isolated objects, like planets, or the trajectory of ideal cannonballs which would be immune to air friction. In this case, a force or a set of forces are applied to the object and modify its motion.

But how can this be applied to a fluid, which is by essence a continuous form of matter? All the small elements of the fluid are in contact and push on each other, forces are transmitted by pressure or contact, and in the fluid there are forces everywhere. It seems necessary to describe all the points of the fluid at the same time, all their velocities and all the forces exerted on them.

$$\frac{\partial \boldsymbol{u}}{\partial t} + (\boldsymbol{u} \cdot \nabla)\boldsymbol{u} = -\frac{\nabla P}{\rho} + \mu \nabla \times (\nabla \times \boldsymbol{u}) + \boldsymbol{F}$$

Physicists in the fields…

A set of points spread in space is what physicists call a "field"; in this way, they speak of a field of velocities, of forces, etc. A "fluid mechanics" equation, like the one above, only expresses Newton's law $F = ma$, but for a field of points, of velocities, of forces. From the fundamental viewpoint, there is nothing new, nothing more than $F = ma$. However, in practice, it is an essential step since real matter is continuous. Gases, liquids which transmit the energy of machines, the air which supports planes, the water which resists the movement of ships, all these fluids do not come to a small number of isolated non-deformable objects like artillery shells or planets. To describe their behavior, we need to know how to apply Newton's law to a continuous set of material points.

That is the reason I have an ambiguous feeling towards this equation (and its cousins): I like most the equations which change my way of seeing the world, but this is not the case here. Despite its high-level appearance, its Greek letters and the "nabla" operator with its esoteric look, the underlying physics is simple — nothing more than Newton. The beauty lies in the application: mathematics allows to orchestrate the motion of all these points in a coherent way, a fluent way indeed. Ultimately, what is more beautiful in an equation: the underlying elementary physics, or the power and scope of its application domain?

We see that "fields" start to appear in this description of continuous matter: the equations of mechanics no longer describe the motion of points, but the variation of fields. This notion will be incredibly fertile in all domains of physics: temperature field, electric and magnetic fields, field of probability, and in the end Quantum Field Theory, where the very concept of a particle will be replaced by the field itself.

An equation difficult to conquer…

Historically, the evolution of "fluid mechanics" is rather parallel to that of thermodynamics, starting from a number of single-case equations, then more and more general ones, accompanied by more precise measurements on the fluids themselves (density, viscosity), and applications to more diverse systems, from rivers to fire hoses to plane wings.

It is even surprising that there are so many applications, although it is in general very difficult to solve the equation! Without a computer, meaning with solely a paper, a pen, and the usual mathematical tools, it is possible to find a solution only for very simple cases, which are in general rather far from practical situations. A stream which runs down a mountain indeed obeys the equation, but it is easy to see that it is hardly possible to describe all the whirls in detail on one notebook page, nor even on many pages… It is even worse, in fact: mathematicians have not yet settled the question whether the equation always possesses solutions. And when it does have one, will the solution found for an initially "quiet" fluid correctly describe its long-term evolution? Or will the modelled behavior become totally erratic? Thus before the advent of computers, physicists and engineers worked hard to find approximations which allowed to get close to practical cases, in order to design complex machines. The engineers of the 19th century and the first half of the 20th century did marvels at designing ships, planes, and even rockets, but they had to include safety margins, and to test small-scale models in the laboratory. In contrast, simple daily-use objects were designed empirically rather than with a calculation.

From small-scale models to supercomputers

When I was a kid, my father told me about the ship model testing pools. Here was a fantastic job, in the eyes of a young boy: I imagined serious engineers who spent long days launching ship models on a pretty basin… The reality was a little less picturesque, and much more technical.

Computers, more precisely the whole development of digital computing, have brought a revolution to the domain. In digital computing, one no longer attempts to solve the equation through formulae, with a paper and pen. A software model of the fluid is built, where the fluid is divided into small volumes, like 3-dimensional pixels. Starting from an initial situation, we simulate the evolution of the fluid for a small step in time, imposing that the equation be approximately verified for this small step. Then we iterate over many steps.

The virtual fluid almost behaves like a real fluid would. Whirls can develop, waves can appear... Of course, there is still much skill involved to perform numerical simulations. For example, smaller pixels and a shorter time-step will provide more accuracy, but will require more operations, hence more time and a more powerful computer. Is the approximate equation used in a step valid in all corners? At all times?

The amazing power of modern computers (a few *billion* operations per second for a workstation, several millions of billions for large computing centers) allows to perform simulations for a wide range of applications. Thus engineers can know how a plane will fly without building any physical model. Almost everything can be tested on virtual models, in conditions much more diverse than with physical models. Wind-tunnel tests are still used, but only to validate the calculations on a small number of configurations, and to study a few cases which are difficult to simulate. With enough computing power, it becomes possible to compute the behavior in real-time, and to build a very realistic flight simulator to test a plane and train the pilots. Digital computing has revolutionized the work of engineers in almost all domains, but maybe fluid mechanics is the most emblematic case. No more bad surprises when the test pilot takes off for the first time!

Chapter 8
Maxwell's equations

$$\vec{\nabla} \cdot \vec{D} = \rho$$

$$\vec{\nabla} \cdot \vec{B} = 0$$

$$\vec{\nabla} \times \vec{H} = \vec{J} + \frac{\partial \vec{D}}{\partial t}$$

$$\vec{\nabla} \times \vec{E} = -\frac{\partial \vec{B}}{\partial t}$$

These are four "coupled" equations, which describe the behavior of electric — $\vec{E}$ and $\vec{D}$, and magnetic — $\vec{B}$ et $\vec{H}$, fields, and electric charges and currents $\vec{J}$ (in passing, they state that *magnetic charge* does not exist). Coupled, because the variables in each of them depend on those in the others, for example a variation of the magnetic field may induce an electric current.

At the origin of this monument of Physics stand several individual laws, Ampere's law, Lenz law, and others…, which each describe one of these "couplings". The power and beauty of Maxwell's equations is to unify these laws in four equations which apply to all electric and magnetic phenomena. They confer the full meaning to the abstract notion of *field*, which we encountered in the previous chapter. As a reminder, a field is an abstract entity which takes on one or several values at every point in space. Take the electric field for example: it can be produced by an electric

$\vec{\nabla} \times \vec{H} = \vec{J} + \frac{\partial \vec{D}}{\partial t}$
$\vec{\nabla} \times \vec{E} = -\frac{\partial \vec{B}}{\partial t}$
$\vec{\nabla} \cdot \vec{D} = \rho$
$\vec{\nabla} \cdot \vec{B} = 0$
MAXWELL

charge, and in a way it represents the influence of this charge at this or that place in space.

Creative equations

Understanding these equations requires some skills in mathematics. One needs to be familiar with the notions of a variation of a field in time and space, the *divergence* or *curl* which are hidden behind the use of the symbol ∇ or "nabla". In any case, these equations are quite elegant, since they simultaneously express the geometric properties of the fields, and their transformations into each other when they evolve with time.

The applications of each single law of induction, charges and currents, are numerous: electrostatic processes, dynamos, electromagnets, electric engines, etc. But Maxwell's equations are for me the first example of *creative* equations: they contain more than what they were written for. Indeed, they describe electromagnetic waves, which Maxwell did not know of, and which would be discovered much later. From radio to UV-light, not forgetting microwaves and visible light, the propagation of all these waves obeys Maxwell's equations. The way they are coupled to charges and electric currents also does: radio emitters and receivers, radar, ionosphere, refraction of light, polarized filters, plasmas, etc.

The Bible of electromagnetism

In short, these equations are both very elegant and very useful. To me, they evoke a certain majesty, first from their status as an intellectual monument, but also from their appearance: four equations like the four façades of a temple, with the very graphical symbols ∂, ∇ decorating friezes and capitals. I remember the feeling of being initiated to electromagnetism as if it were a religion, with Maxwell's equations standing as Tables of the Law.

The cult: first the first exercises, where we try to apply the equations to simple cases, for example to recover Ampere's or Lenz laws, elementary results. Then more complex applications, further requiring the power of the equations, the complex cooperation of all the terms…

And of course this religion had its book: *Classical Electrodynamics* by John David Jackson. First issued in 1962, then reprinted twice with more and more chapters and exercises, the "Jackson" is the Bible of Electromagnetism. Even more, since it contains the foundations (the Old Testament), the education of the faithful (the New Testament), and even many esoteric developments for a sect of enlightened (The Kabbalah).

The "Jackson" is not an easy access book. It is reserved to courageous students/monks who accept to invest their time and to devote a lot of effort to it. Even before starting its study, the book is defended by a formal rampart, because it uses an obsolete system of units ("cgs" notation) instead of the international system (SI): meter, second, ampere, volt, etc. Hence all the equations differ from their usual form.

Admittedly, the more recent editions comprise a double page which is supposed to ease the conversion between the cgs and SI systems. Since I have often tried to translate my calculations from one system to the other, I now think that this "conversion" is but another ordeal, much like the tonsure of the novice. After having obtained for the calculation of the magnetic field produced by a flashlight battery and three turns of copper wire, a value higher than that reached close to a spinning black hole, one suddenly becomes humbler.

At the end of each chapter, the "Jackson" also proposes exercises. According to legend, some humans have done them all. I guess that to recognize each other they use a secret sign, like some way of placing their fingers during an introductive handshake, or a surreptitious exchange of passwords during a trivial conversation. The sign is probably revealed by the completion of the dreadful calculation of the last exercise of one of the chapters. Or maybe the one before last: it would be too easy to do all the last ones, all very difficult… I have not done them all, and I did not receive any revelation. I have met some fabulous old-school physicists, and I was thinking: maybe him, or her? But how could one ask, without humiliating the other person, or humiliating oneself: "And you, the exercises in the Jackson, did you ...?"

In principle, the Jackson only deals with the classical theory of electromagnetism, not the quantum theory. The fields, objects and equations keep the full rigor and purity of Maxwell's equations, and do not condescend to that quantum world with fuzzy contours and uncertain predictions. Or does it? In several occasions, it ventures to the boundary, brilliantly distinguishing what is truly quantum from what is not, classically recovering some results that one always imagined as quantum. Real Art, like finding Velasquez behind Dali or Shakespeare behind Beckett…

A mischievous "Z^0"

In 1983, I was a student on the UA2 experiment, which studied the proton-antiproton collisions produced by the SppS collider at CERN. UA2, like the competing experiment UA1, was searching for the Z^0 particle, the star of that time, whose existence had been predicted by theory. In particular, it could decay to an electron-positron pair, two particles that our detectors could identify and measure. All that was left for us was to find it…

Among all the collisions, the UA2 suddenly detected an event which contained an electron and a positron… but accompanied by a high-energy photon, well separated from the other two particles. Immediately, the whole "collaboration" — the team in charge of the UA2 experiment — was swarming with excitement. Was this already a Z^0? But then what was this extra photon? Could it be the sign of an exotic, unexpected particle, like an excited electron? Or was this a normal Z^0, with inner radiation, an effect known in the decay of other particles? In a word, what was the probability that a *bona-fide* Z^0 would decay this way, into an electron, a positron, and a photon?

Nobody was expecting this question, or had the calculation already at hand. Admittedly, Quantum Field Theory allowed to calculate this probability, but we did not have the programs we have today, which can make this kind of calculation almost automatically. We had to carry it out ourselves. The calculation is a bit complex, and it demands some practice to be sure not to forget a factor of 2 or π somewhere. But one of us (me or my thesis supervisor, I am not sure now) remembered having read in the Jackson a paragraph on a very similar topic. The author was even emphasizing that the result was universal, and did not depend on the details of the process.

Adapting the calculation in the Jackson to our case did not take long, the longest being, as you could have guessed, to find our way through its damned units. The result was surprising: this kind of event was not improbable, with a probability of occurrence of a few percent. The most likely hypothesis was that we had indeed discovered our first Z^0, but that it had played a small trick on us. The confirmation of this probability by the real Field Theory calculation came a few days later, then in the next months the experiment observed a handful of plain good-looking Z^0 events. The competing experiment also did, and the whole thing brought the Nobel Prize to Carlo Rubbia and Simon Van der Meer, but this is another story.

Jackson himself (who died in 2016), was an excellent physicist beyond his book, but he did not receive the Nobel Prize, and indeed nothing in his book is a revolution in Physics. Nevertheless, I think that here is the genius of a top-of-the-art craftsman. Maxwell equations are gold, and Jackson was and still is their best goldsmith.

Chapter 9
The matter-energy equivalence

$$E = mc^2$$

On the left-hand side: energy, on the right: mass, multiplied by the velocity of light squared. This equation expresses the equivalence between mass and energy — we will see in which context a bit more in detail — the touchstone of Einstein's relativity. A theory that revolutionized the way we see matter, space, and time.

Between respect and awe

Why did such a simple equation become the very symbol of Physics?

The good reasons: the personality of its inventor, his clear-sightedness, the scientific revolution that is the theory of relativity.

But above all, the bad reason: it is impossible to ignore that this equation allowed to understand the formidable energy at play in nuclear reactions, and had as a first consequence the creation of the atomic bomb.

Thus, $E = mc^2$ fills us with respect mixed with awe, very much like the fear of the divine in the Ancient Testament. The contrast with the popular image of Albert Einstein, a rather positive image, increases the feeling of anxiety. How could this good-natured, almost facetious, grandfather be at the origin of the sacred fire, offering to a few the possibility of taking the life of many humans in a single move?

Strange relations between space and time

When Einstein formulated the equation, in 1905, at only 26, he was a plain but good-looking young man. The messy hair would appear in the 1920's, the

$E = mc^2$

"friendly mad scientist" look in the 1930's only. Still very far away from the now classic shortcut: messy hair = mad scientist = irresponsible genius = consequences as immense as unpredictable.

In this very early part of the 20th century, radioactivity had already been discovered ten years earlier, and on the theoretical side, the field had been considerably prepared. In fact, Maxwell's equations (chapter 8) unknowingly contain the essence of relativity, in particular the fact that the velocity of light in vacuum is the ultimate velocity, independently of the frame in which it is measured. The theorist Hendrik Lorentz could read them, and he deduced his formulae known as "Lorentz transforms" which enunciate strange relations between space and time.

Henri Poincaré notably went down this road, but he was blocked by the wall of intuition. This great scientist had already underlined that everything pointed to, when travelling at a high speed, time slowing down and distances shrinking. But how was this possible? Poincaré put the problem in perspective perfectly, but he stopped there, thinking that reality must be different, that this could only be a calculation artefact.

To believe, truly…

At this stage, Einstein's genius was simply to *really* believe in the theory. All the formalism was there, it sufficed to say that it *truly* described reality.

One small step… The immediate consequence was $E = mc^2$. When he found his equation, was Einstein already searching for an explanation to radioactivity, or did he realize only after? Was he only trying to extend the work of Lorenz and Poincaré? Was this the outcome of his own reflection on space and time? Was it important that he was in charge of the synchronization of the Swiss railways?

Indeed, is the "true" inventor the person who constructs the model (the theory) without believing in it, or that who first believes in it? One starts by building a model, only to describe experimental facts, then suddenly this model yields a completely different understanding of the world around us.

Modern Physics offers several cases of such a tipping point: Max Planck and his constant, Einstein and relativity, Dirac and his equation as we will see later in this book. These "paradigm shifts" pertain to the Philosophy of Science, but to a mere experimental physicist like me, knowing these great examples is an incentive to keeping an open, flexible mind.

In day-to-day work, a scientist does not invent a new theory of the world, and one does not discover a bright new unexpected effect every week. But we do look at all experimental and theoretical results with curiosity, with sharpness: how does this new result integrate with what I already know? Is this effect important or secondary? Most of us will not be Einsteins or Diracs, but even at our modest level, what counts is to have one's own model of the world, one's own consistent personal construction, with which it is possible to challenge one's own results or the results of others. Again: to know what we know, and what we don't know.

The energy of the sun...

$E = mc^2$ unveils the formidable energy (formidable in those times, and at the scale of a particle) at play in radioactive disintegrations. A particle with mass M decays into two particles with masses m_1 and m_2, with the sum of the masses $m_1 + m_2$ smaller than M. The difference in mass between the initial state M and the final state $m_1 + m_2$ is converted into energy:

$$E = [M - (m_1 + m_2)]c^2$$

c is a very large number, hence this energy released is very large.

One of the most immediate and striking consequences is that at last one understands why the Sun shines. Not yet in detail — this will require the evolution of Nuclear Physics — but it unveils the possibility of a new source of energy, which could allow our star to produce a huge amount of energy, and for a very long time!

Around the end of the 19th century, considering the known sources of energy (chemical combustion, gravitational collapse), the age of the Sun could be estimated to at most a few million years. It was indeed the topic of a much wider controversy on the age of the Solar System and of the Earth. Darwin had observed fossils and geology, and he deduced an age of at least several hundred million years, sufficient to allow the natural evolution of species. But physicists (among which Lord Kelvin) claimed that such a duration of combustion of the Sun was impossible. Thus, at that time, physicists *opposed* Darwinism, a rather paradoxical situation as seen from today!

As soon as Henri Becquerel discovered radioactivity, a hypothesis was put forward: this could contribute to the energy produced by stars, including the Sun. But it is indeed $E = mc^2$ that brought a satisfactory explanation. After a few advances in Nuclear Physics, the mechanism of nuclear fusion was elucidated. It even became quantitative, and in 1920, Sir Arthur Eddington, made the following statement:

"If, indeed, the subatomic energy in the stars is being freely used to maintain their great furnaces, it seems to bring a little nearer to fulfilment our dream of controlling this latent power for the well-being of the human race — or for its suicide".

… and of the atomic bomb

The equation $E = mc^2$ remained limited to theory and experiment of fundamental research on particles until 1938, when Otto Hahn and Lise Meitner[1] (with the help of Fritz Strassman and Otto Frisch) discovered nuclear fission. Then appeared the possibility of a chain reaction which could involve large quantities of atomic nuclei, like several kilograms, releasing enormous amounts of energy. Furthermore, this was no longer the domain of theory or of stars, since such a reaction could be put to work by man, starting from natural elements like Uranium, liberating gigantic energy from a tiny amount of matter.

To set the scale, if it became possible to release it, the energy available in a piece of matter is about a million times larger than the energy which could be released by the ordinary (chemical) combustion of the same amount of matter. The possibility of a weapon was pointed out almost immediately.

[1] Only Otto Hahn received the Nobel Prize in chemistry, 1944. Not Lise Meitner, who first really understood the mechanism of nuclear fission, one of the worst examples of discrimination…

Is it a coincidence that this discovery was made at the very moment of extraordinarily troubled political times, at the very place of the trouble and at its cultural heart, since many actors of the discovery, like Albert Einstein and Lise Meitner, were among the most threatened? Can we think that reason and morality happened (this time) to work together to defeat obscurantism and barbarism? By the way, was it possible that in other occasions, they conspired to tip the scale on the *wrong* side, and if yes, when?

The rest is History: Einstein's call to Roosevelt to convince him to build the "Atomic Bomb" before the Nazi Germany, the ultra-fast development of the "Manhattan Project" in the USA, and of course the dreadful final outcome at Hiroshima and Nagasaki. During the first nuclear test (the *Trinity* test), Oppenheimer, one of the most involved physicists in the American making of the bomb, remembered an excerpt from the *Bhagavad Gita*: "brighter than a thousand suns". That said it all.

This story has marked $E = mc^2$ forever, and with it all nuclear science, and in turn all science. Standing as an original sin, it allows for all fears, all fantasies, and all kinds of manipulations. We should learn to regard it with some distance, if we want to dispassionately confront the new questions about scientific ethics. In any case, the story should be included in the minimal education of every physicist, and why not, of every scientist.

Chapter 10
Schrödinger's equation

$$H(t)\left|\psi(t)\right\rangle = i\hbar\frac{d}{dt}\left|\psi(t)\right\rangle$$

Quantum Mechanics… How many students have dreamed about these words, including myself of course. The key to an uncertain, poetic world, full of chances and fluctuations… Full of mystery too: particles travelling several paths at the same time, and which are never stopped by any wall; and of course, cats that are simultaneously alive and dead. I always hated this image of Schrödinger's poor old cat. What a strange idea you had, Erwin, you could have chosen a less irritating example. In some textbooks one could see the superimposed drawings of a vivid cat and a pitiful corpse. In my public conferences, I replace this awful image by that of the Cheshire Cat who, in *Alice in Wonderland*, appears or disappears, leaving only her smile behind. A very quantum cat indeed, and so long before the concept existed…

Schrödinger's dead-alive cat is still a substantial line of questioning for physicists, and the object of many fantasies in the public eye. But the Austrian physicist's equation has drastically changed the way we see matter, and the world. For sure, like all equations and theories, it was not born from nothing. Before it was formulated, Planck, Einstein, and Louis de Broglie had already interpreted — at the beginning of the 20th century — experimental observations in terms of "quantum mechanics", but it was rather a set of rules allowing to describe the results, not yet a different vision of matter.

$H(t)|\psi(t)\rangle = i\hbar\frac{d}{dt}|\psi(t)\rangle$

“Psy” or “Psi”?

Schrödinger’s equation is indeed the first to claim loud and clear, that matter behaves in a way completely different from what our senses indicate to us. And amazingly so! Hang on: physical objects are not particles, but “wave functions” (the ψ in the equation). In practice, these are complex numbers whose modulus (their “size”) indicate the probability of finding a particle here rather than there. The energy is no longer a number, but a mathematical operator (H in the equation) which somehow chooses the result when we undertake a measurement. Add to that Planck’s mysterious constant h, and an “i” for “imaginary” (the famous complex number such that $i^2 = -1$, well known to confuse high-school students) and you have all the ingredients required for a mythical equation, almost philosophical. Not surprisingly, Schrödinger himself became an adept of Vedic studies!

By the way, the wave function is represented by ψ, which we read “psi”! A reckless notation, which will encourage a dangerous collision between quantum mechanics and the psyche. It is true that the observer plays a central role in quantum mechanics, since it is the observation of a measurement which determines the state of the system. (Wo)man and consciousness are indeed at the center of physics, and all this is the source of big and beautiful questions, which go far beyond this text.

But the naïve similarity between ψ and “psy” has also allowed for all kinds of excesses and misuses. Thus, we have seen poor quantum theory milked for any bad use: paranormal, “quantum Body”, “quantum psychology or therapy” — a disciple of Lacan even wrote “kantum (*sic*) formulae of sexuation”! Then everything becomes possible, by the operation of the “psi” or of the “psy”: bending teaspoons by the power of thought, healing the worst diseases, etc.

I must admit that physicists were sometimes deliberate accomplices of these missteps, more by humor than by malice. For example, among professionals, they speak very seriously about quantum “teleportation”. Maybe they should explain to the public that this is not about spending a weekend at the other end of the galaxy, but about recreating a *very special* state of a *very simple* system by only transmitting information. For sure, they could have chosen a different term, but this one was just too tempting!

The first quantum vertigo

In fact, there is no need for any ambiguous "psis" or "psys", or teleportation problems, to feel the quantum vertigo. Students first learn Schrödinger's equation at university. As (at least in France), one does not learn much about the History of Sciences, and there is little time to see all the experiments and the line of reasoning which preceded it, students may wonder how Schrödinger came to it. But the equation itself is rather simple, hence we can quickly go to the first exercise, the first application.

Question no 1. A small well is drilled in some material, and a particle (or a small ball) is deposited at its bottom. Calculate the motion of particle.

Once the variables are properly defined, solving the equation is elementary. The result is a tad strange, but not alarmingly so: there is no solution with a motionless particle. The smallest allowed movement is not zero, but rather a light oscillation at the bottom of the well.

Then follows another surprise: if we consider slightly larger movements it is not possible to choose the amplitude arbitrarily, since it must be a multiple of the minimal amplitude: two times, three times, etc.

A bit strange indeed, but the title of the course said it: this is *quantum* mechanics, you could expect a few oddities. Nothing revolutionary though: the amplitudes of motion are *quantized* instead of being continuously variable like in classical mechanics. Not yet the big thrill the students expected! But this is reassuring: it seems rather familiar, not so far from the "classical" situation…

Question no 2. Besides the first well, a second well is drilled, then the particle is deposited at the bottom of the first well, like at question no 1. Calculate the motion of the particle. The resolution is only slightly more involved, and the solution is a formula which looks a bit complicated. Then the exercise goes on: derive the probability of finding the particle in the *second well.* We compute carefully, and we find that… this probability is not zero. We go again through the calculation to check that we did not forget a term or a sign, but no: by putting a particle in the hole (1), there is a non-zero probability to find it in the hole (2).

Simple and clear. There is no "classical" analogy, no intuition to hang on to: welcome to the quantum world!

Such a counter-intuitive world

At this stage, we can only halt for a while and try to come to terms with reality being quantum. We have to build up new intuitions. And this is hard, because this reality does not at all correspond to that which our senses feel. So much so that years of practice are necessary simply not to tell too much nonsense...

You must have read some quizzes proposed by science magazines. Their goal is to reassure the reader about her level of understanding, with very direct questions like the following one. A boat floats in a swimming pool, with a man and a big dense stone aboard. The level of water of the pool is noted. The man throws the stone into the pool. After stabilization, the level of water is again noted. Is it higher of lower than before the stone was thrown? For a physicist, this kind of exercise is really easy, but this would not be the case for its quantum equivalent...

I am not a specialist of all the subtleties of Quantum Mechanics, but I have studied it rather deeply. I constantly use it in my everyday work on particles. However, if an expert would offer me such a test in its quantum version, I would never be sure of the correct answer (before undertaking the real calculation of course). Admittedly, with daily practice, my life takes place in part in the quantum world so to say. But it is easy to slip: the power of the intuition dictated by our senses is irresistible, and our common sense is all but quantum.

A wealth of applications

After this first quantum vertigo, comes the light. In class we learn how to solve Schrödinger's equation for the motion of the electrons of an atom. This opens an immense field: suddenly, we understand why atoms emit or absorb light at particular wavelengths. Immediately, we see for example that the light emitted by a heated atom contains the signature of this atom. In turn it will be possible to identify the chemical components of matter at a distance, in the lab or in distant stars. Even better: with the Doppler effect, it will be possible to measure the velocity of this atom with respect to us, even at the other end of the Universe.

Then come the countless practical applications, solid state physics, semiconductors, the transistor, electronics, the extraordinary development of micro-electronics, which now packs in a smartphone a computer more powerful than a whole computing room of the 1970's.

All this is practical, robust, useful. People talk to each other from continent to continent, screens inform and entertain, markets go up and down at light speed for better and for worse — rather for worse in the latter case.

And yet, the paradoxes of Quantum Mechanics are still with us. Most physicists shyly turn away, when others, the specialists, design wonderful experiments to understand if it really, *really* works this way. The result is "yes", each and every time. Schrödinger is right and our senses deceive us. And his cat, alive indeed, displays her most beautiful smile, or smirk… A real cat's smile…

Chapter 11

Heisenberg's uncertainty relations

$$\Delta x \cdot \Delta p_x \geq \frac{\hbar}{2}$$

With this formula we are at the heart of Quantum Mechanics. The message of this formula is very simple and at the same time it is absolutely incredible: all notions of particles, positions, trajectories, velocities are questioned. Thus I will try to explain it in some detail. Don't be afraid, the mathematics will remain elementary: only a few multiplications.

You have noticed it immediately: the usual sign " = " of an equation is replaced here by " ≥ ", the mathematical symbol for "larger than", "superior to". This is an inequality, which means that the formula will not be used directly to calculate a motion or the size of a process.

At the heart of the quantum fuzziness

This inequality is not a physical law in the sense of an operational rule. It rather describes a consequence of the laws, here the rules of Quantum Mechanics. The subject of the inequality is the simplest system one can imagine: the motion of a single object. "x" represents the object position, as measured along a ruler for example, in meters. The symbol "Δx" describes the *uncertainty* on the measurement of x, namely the size of the interval in which I can claim the object is located. For instance, writing $\Delta x = 1\ mm$ means that I can measure the position to a 1 mm accuracy.

$\Delta x \cdot \Delta p_x \geq \frac{\hbar}{2}$

"p_x" is the momentum, which is simply the product of the mass by the velocity[1]:

$$p_x = mv_x$$

Similarly, to the position, Δp_x is the accuracy on the momentum; as the mass m is constant during the motion, the uncertainty on the momentum is simply related to the uncertainty on the velocity:

$$\Delta p_x = m\Delta v_x$$

Thus, $\Delta v_x = 1$ meter/second (m/s) if[2] we measure the velocity of the object with an accuracy of 1 m/s.

In the right-hand side term of the inequality, h is Planck's constant, a universal quantity, just like the speed of light in vacuum or the constant of gravitation. Then the formula states that the product of the uncertainties with which I can define the position and the velocity is always larger than a fixed quantity, independent of the movement.

An imperfect world...

I am a physicist, and since I am proud of my scientific super-powers, I have built a fantastic detector. I am ready to measure the position of a particle with perfect accuracy. A perfect accuracy should be translated into an uncertainty interval reduced to nothing: $\Delta x = 0$ *mm*.

Fine, but in this case, the left-hand side becomes 0 times Δp_x, i.e. 0. It is impossible that 0 be larger than the quantity on the right, which is not null. Hence it is impossible to determine the position with a perfect accuracy. What about the velocity? Same reasoning: if the precision on the velocity is perfect, the value of Δv_x is 0, and in turn the left-hand side is 0, which is again impossible. In this new mechanics, it is forbidden to measure the position or the velocity with perfect accuracy!

---

[1] When the velocity is much smaller that the speed of light; otherwise the formula is a bit more complicated, but this does not change anything in substance.

[2] 1 m/s is equal to 3.6 km/h.

But our everyday world remains sharp!

Yet, in the world surrounding us, the positions of the objects do not seem to be submitted to any uncertainty. Nor are the velocities: it would be difficult to invoke Quantum Mechanics to contest the measurement of a speed camera on the road!

Here is the trick: the quantity $\hbar$ is extremely small, so much so that the equation does not limit the measurement accuracy for ordinary objects. For example, let us consider a 1-gram marble, for which we want to measure the position to a $1/10^{\text{th}}$ of a mm, and the speed to 1 mm/s, already very good accuracies to track the motion of such an object. The product $\Delta x \; m \; \Delta v_x$ is still 10^{24} (hundred thousand billions of billions) times larger than the minimum $\hbar/2$. Quite a comfortable margin!

In contrast, atoms, atomic nuclei, and elementary particles constantly evolve at the edge of this limit. Position and speed become fuzzy concepts, and can only be determined statistically, for instance by taking an average over a large number of measurements.

A perfectly pure randomness

Even stranger: not only measurements, but also events are submitted to the uncertainty, as shown by radioactivity, discovered by Henri Becquerel in 1896. Some atoms, natural or artificial, are radioactive, meaning that their nucleus can spontaneously disintegrate, producing different smaller nuclei, and a few other particles. Considering a large quantity of such atoms, each second we see a certain fraction of them decay. But if we study a single atom, we realize that, due to Heisenberg's uncertainty relation, it is impossible to know *when* it will disintegrate.

This is frustrating, inasmuch our intuition, built on our senses, demands to find a cause, a mechanism for any event we observe. However, there is *no* mechanism. After Quantum Mechanics was invented, scientists long sought the hidden mechanism which would control the decay of a radioactive atom. One could imagine that each atom was programmed at the time of its formation to decay at such a date, or that a wear and tear mechanism degrades the condition of the atom until it finally decays, like a worn-out electric bulb.

In a sense this would be reassuring, but it is not the case. Many experiments have demonstrated that the advent of the disintegration is purely random: the atom is absolutely like new… until it suddenly decays! If it has not done so yet, the probability that it decays in the next second is exactly the same as the probability it had to decay in a second's time a billion years ago.

An operational "mechanics" but a mysterious one

Yet, the laws of Quantum Mechanics are perfectly rigorous. In particular, they allow to predict, with extraordinary accuracy, the *relevant* quantities for elementary particles: energy levels, probability of localization, of interaction, etc. Practical applications are already numerous: behavior of semiconducting materials for electronics, lasers, structure of molecules, chemistry and biochemistry, and this is only a start. Scientists are now looking for a quantum computer, where this fuzziness would be put to good use, such that a very complex computation would no longer be carried out step by step by a processor, but rather in parallel by a large number of particles (electrons or photons).

There is no doubt that Quantum Mechanics is operational, even if we still have not unraveled all its mysteries. This uncertainty on the measurement of intuitive quantities, like position, speed, duration, has deeply troubled scientists and philosophers. Scientists have sought "hidden variables", which would allow for holding onto the intuitive scheme of cause and effect. Not only were none found, but beautiful experiments were conceived in the 1960's (first by John Bell), then performed in the 1980's by the pioneer Alain Aspect, followed by many others. The results were crystal clear: there are no hidden variables, and despite its baroque behavior (particles follow two paths at the same time, or seem to know in advance what we will do before we even decided it!), Quantum Mechanics always predicts the correct result.

Then physicists need not worry, and the most frequent attitude is to ignore the problem of the hidden meaning, to only see the astounding efficiency of the theory for describing the world around us. However, the questioning on the substance is always present, and constitutes a source of intellectual and practical stimulation. The imagination needed to stress the theory is a nice example of creativity, rigor and of experimental skills (I say it very freely since this is not my own experimental field!).

Progress, an obsolete concept?

Some have seen in the quantum uncertainty the illustration of a fundamental change of status for science. The second example would be the discovery of chaos which we mentioned above about gravitation, and the third, Gödel's incompleteness theorem.

In Quantum Mechanics, Heisenberg's relations forbid the prediction of the speed and position of particles without uncertainty. When Newton's gravitation is applied to three bodies or more, it gives chaotic solutions, which make the prediction of the position of planets in a few tens of million years impossible in practice. Finally, Kurt Gödel demonstrated around 1930 that simple mathematics (like arithmetic) contain undecidable propositions, which cannot be proven true nor false.

These concepts have something in common indeed: elaborated at the turn of the 19th and 20th centuries, they all seem to point to intrinsic limits of scientific knowledge. In the 1970's, some intellectuals had integrated these confessions by science itself, into the large "postmodern" movement of doubt on the legitimacy of science and progress.

Of course, these doubts and questionings have more immediate sources, more accessible to the public, and quite concrete, like the role of science in nuclear weapons, environmental destruction, consumer society, etc. But the fact that science itself admitted its own limits played a role in the intellectual and power circles. Taken out of their context, the words "uncertainty", "chaos", "undecidable", became symbols of a powerless, useless and even aberrant science.

I discovered this in 1996, at a time when I was a physicist deeply absorbed in my domain, and not much involved in outreach. Like most of my colleagues, I had little time to go looking around, but I already had some curiosity for epistemology, in particular about the interpretation of quantum mechanics, which was the topic for long discussions with my old friends Hervé[3] and Jean. During the summer of 1996, the newspaper "Le Monde"[4] ran an important series of articles with the title "Progress, an obsolete concept?". I had already heard such messages, notably in the magazine "La Recherche"[5], which in those times offered a large stage to postmodern philosophers. I had first deemed this only as a debate between professional philosophers, besides, I had no time to pay attention to it. But when the daily newspaper of reference questioned the very idea of progress, it shook me up enough that I decided to dig more into the subject.

I then discovered "postmodern" epistemology and its slogan *Science is a Social Construct*, a shock to a rational scientist, brought up in the cult of Enlightenment and Progress. I am not competent for the sociological and philosophical part, but on this occasion I also observed the strange association of scientific concepts such as "uncertainty", "chaos", "undecidable" with this line of thought.

This hodgepodge seemed to me to be the exact symptom of what these philosophers wanted to denounce: an elitist society where admitting one's limits is interpreted as a weakness, or even a fault. Really? After two centuries of linear progress, without much questioning, around the end of the 19th century science reached a level of awareness high enough to ask questions about its tools (mathematics), its methods (calculus), and its interpretation of the world (objects). Is this a weakness? Should this challenge the notion of progress? I think not; on the contrary this shows a certain maturity, or even a true wisdom.

If they are not taken out of their context too quickly, "uncertainty", "chaos", and "undecidable" only show us that nature (whatever we put in this word, an objective reality or not), is not as our senses directly tell us. Other interpretations may be considered, and the notions of position, velocity, event, are not adapted beyond our surrounding environment. Other

[3] Hervé Zwirn has written several books on this subject.

[4] Le Monde is a moderate-left French daily newspaper with an intellectual ambition. Summer in France is rather empty for politics, hence it is an opportunity for deeper studies...

[5] La Recherche is a French monthly science magazine, comparable to "Scientific American".

measurements are possible, of probability, energy, correlations, other modes of reasoning, and finally other forms of progress. To associate these discoveries to a possible downgrading of science, amounts to denying a person the right to halt to think and to accuse him or her of weakness as soon as (s)he expresses a doubt.

Chapter 12

Einstein's equations, general relativity

$$R_{\mu\nu} - \frac{1}{2} g_{\mu\nu} R = \frac{8\pi G}{c^4} T_{\mu\nu}$$

The large capital letters "*R*", "*G*", "*T*" are conclusive: here we are no longer in the fluctuating quantum states of ψ or h, but in the rigid and eternal geometry. The indices μ, ν, placed in pairs at the foot of the large symbols, are no longer appoggiaturas, but milestones on the grid which measures space and time.

Rigid? Not that much. The equation rather describes how space-time deforms under the action of matter: in (very) short,

$$R_{\mu\nu} - \frac{1}{2} g_{\mu\nu} R$$

represents the qualities of the space-time grid: the size of its mesh, its curvature, while

$$T_{\mu\nu}$$

represents the presence of matter or energy on that grid.

Dominating effects for cosmology

Indeed, since 1917, and the invention of General Relativity by Albert Einstein, space (space-time in fact) is no longer a uniform and unchanging grid. It is still a grid, but deformed, compressed, inflated, curved, bent, as soon as matter is present. These distortions, these curvatures modify the

$R_{\mu\nu} - \frac{1}{2}g_{\mu\nu}R = \frac{8\pi G}{c^4}T_{\mu\nu}$

trajectory of celestial (and terrestrial!) objects: we interpret this as a force, which we call "Gravitation". Newton's equation used to say:

"Objects with mass attract each other"

Einstein brings the following correction:

"Objects with mass distort space-time around them. Other objects feel this distortion, and the whole interaction appears like an attraction."

In our everyday environment, the solutions of Einstein's equations differ very little from those of Newton's equation, and apples do fall where they are expected. The only (but well-known) example of a practical consequence is the implementation of the GPS system: the position of a GPS receiver on Earth is computed from the measurement of travel times of radio signals between several satellites and the receiver. These are very accurate measurements between moving (and rotating) objects, furthermore in Earth's gravitational field. Relativistic effects then become quite visible, and they were accounted for when the system was designed.

Even at the scale of the Solar System, the change is very small: a tiny deviation of light rays, a minuscule advance of Mercury on its orbit...

The power of General Relativity only appears with very large scales. Relating time, space, and matter, it gives to physicists the means of questioning the geometry and the content of the whole Universe. Modern cosmology was born out of this equation and of the development of observation devices.

Describing the whole universe

Throughout the first years of university, I had read various articles and books for the general public about General Relativity, but I only had a course on this subject in the very last year of the master's degree. This first real contact made a strong impression on me, and I remember a feeling of respect towards the rigor of the geometry, the immensity of distances and forces at play, the smooth, continuous character of the equation...

In essence, this is no more than what everybody can feel when looking up at a sky full of stars. Pascal confessed: "The eternal silence of these infinite spaces scares me", and, believe me, mastering Einstein's equation does not bring any reassurance! Raising one's nose, one can indeed be seized by awe, or just be satisfied by some familiarity, for example by recognizing the constellations, the usual planets, etc. Learning cosmology amounts to learning the *reality* of gigantic distances and time-intervals, and to feeling in a quantitative, sensible, personal way the size and place of humans in the Universe.

Supported by General Relativity, and thanks to a long chain of observations and astronomical measurements from Earth and from satellites, cosmology has the ambition of describing the shape and evolution of the entire Universe. This is a revolution, which started at the end of the 1920's, when Canon Lemaître proposed a solution to the equations which predicted a Universe in continuous expansion, apparently confirmed by Edwin Hubble's measurements. Until that time, the Universe was believed to be eternal and unchanging; suddenly we discovered that it is likely not the case.

Expansion and big bang

This means that two distant points in the Universe, say two galaxies, move away from each other. Let me just correct, if needed, the idea that expansion would mean "explosion in space from a tiny volume". No, it is quite possible to have an infinite size and at the same time to undergo an expansion. To visualize this, you just need to figure an infinite straight line, with no limit on the left or on the right. We can move all points away from each other by multiplying all distances by a factor of two, in a way stretching the line into itself. No problem since it has no extremities... This is exactly what happens for the Universe, but in the three dimensions of space (length, width, height). It is strange enough that something so easy to figure out for a straight line, is more difficult for three dimensional space!

Therefore, the present Universe is expanding. And as it behaves more or less like a gas, it cools down. But what happens if we run the movie backwards, moving back in time from today? We see the Universe contracting and warming up. We could imagine that by pursuing this process, its density and its temperature would smoothly increase, and become infinite at an infinite time back in the past. But this is not what the equations say: by going back in time, density and temperature increase faster and faster, so much so that they both become infinite for a definite date in the past. More precisely, this date is 13.8 billion years ago with the present estimation. This is the famous Big Bang model.

The evolution of the universe

After the advent of General Relativity, and thanks to the development of Nuclear Physics, the functioning of stars and their energy production were elucidated. Scientists also understood that almost all chemical elements around us today were not present at the beginning of the Universe, but were rather produced by nuclear reactions during the evolution of stars or during their explosion, like supernovae. This brings a very concrete meaning to the expression "we are all stardust" ... Only a few "primordial" elements were produced in the very first minutes of the Universe, and the Big Bang model is able to describe the conditions of the synthesis of these first atomic nuclei. The measurement of the relative abundance of these elements in the cosmos provided, around the middle of the 20th century, a shining (indeed!) confirmation of the Big Bang model. Since that time, numerous observations in various domains have put the model to test, and it has resisted perfectly until today.

If we put aside the creationist thesis, and we rely only on observations, the very idea of an "age of the Universe" is a complete revolution. In the human experience, what is more permanent than the motion of planets and stars? The repetition of seasons, of the position of planets, of eclipses, everything shows to humans that their life span is but a tiny interval in an unchanging Universe. Agreed, our life is very short compared to the 13.8 billion years of the Universe, but we have understood that it has not always been there. And we have been able to understand that by only observing, and thinking. Again, we freed ourselves from our senses and our intuition...

Black holes: A laboratory... in thinking

Thanks to modern observatories, the effects of General Relativity can now be seen in a spectacular fashion: light rays from distant galaxies, following the space-time grid distorted by the presence of matter, display gravitational mirages, huge luminous arcs or multiple images. Recently, using an extraordinarily sensitive apparatus, physicists have detected the passage on Earth of gravitational waves emitted by the fusion of two very distant black holes. This observation is again perfectly compatible with General Relativity.

Well known to the public, black holes are a direct consequence of Einstein's equation. Around their formidable concentration of mass, space-time is so curved that light-rays cannot escape. It now seems that black holes are everywhere, starting with the center of our galaxy, the Milky Way, where sits an "average" sized black hole of nearly 4 million solar masses — to be compared to the 17 billion solar masses of the most gigantic black hole known today…

Black holes are extraordinary astrophysical objects, but they are also a wonderful investigation area for general physics. In this extremely compact world, closed upon itself, all the laws are challenged. Even the concept of *information*: when objects, or particles, fall into a black hole, what happens with the information contained in their nature, their energy, their direction?

This mystery still puzzles scientists, even more so because Stephen Hawking showed that black holes slowly evaporate… Once the black hole has totally evaporated, where has the information, which was carried by all the particles it had absorbed, disappeared to? A crucial point is that Hawking's evaporation is a pure quantum effect, and the theory of General Relativity is terribly at odds with Quantum Mechanics. A black hole is really the best place to confront, presently only in thought, two visions of the world so different: General Relativity, ideal, smooth, geometric, and Quantum Mechanics, strange, fuzzy, fluctuating…

Here and now

The most revolutionary aspect is perhaps the possibility, or at least the intent, to consider the whole Universe as an object for measurements. Maybe this is only another vanity, and humans in the future will in turn regard us as we regard our ancestors from the Middle Ages, who thought they inhabited a flat Earth under a velvet canopy sprinkled with crystal stars.

Maybe, but let us be carried by this Promethean ambition: thanks to observations, and their interpretation in the framework of General Relativity, the Universe now has an age (13.8 billion years), a shape (flat, no curvature), a density (10^{-26} kg/m^3, equivalent to about 6 proton mass per cubic meter, an easily remembered number!). The model of the Big Bang allows to describe its evolution, from a few nanoseconds after its birth down to today.

The Universe as a measurable object is also an open window to all kinds of creativity. According to some theoretical models, the Universe would be twisted, or fold onto itself, some configurations that no Science-Fiction author would have dared proposing! However, even these models were confronted to observations, and it seems very likely that our Universe is as simple and as flat as possible.

Then that's what this is all about? An infinite Universe, flat, which slowly gets more diluted and cooler? Some scientists are a bit disappointed. Instead, we could be happy to be "lucky" enough that we are at an era when the forces which drive its dynamics balance each other harmoniously, and when it still contains enough structures to host a comfortable galaxy, in the suburbs of which we live, safely remote from the monstrous central black hole…

Chapter 13
Dirac's equation

$$(i\gamma^\mu\partial_\mu - m)\psi = 0$$

It still stirs me like a child... The most beautiful, the purest of all. Much more beautiful than $E = mc^2$, and not stained by a cruel History.

Even if one does not like mathematics, even if the rationality can inspire some rejection, one can only admire the elegance of the characters: the harmonious roundness of the γ, the gentleness of the ∂, the sharpness of the first i, the delicateness of the μ indices set like appoggiaturas, and the deep mystery of the ψ.

I admit it is a bit confidential, since it is only taught in the last years of university, requiring a strong background in physics and mathematics, namely Relativity and Quantum Mechanics.

And Dirac created antimatter

As we shall see, Paul Dirac could read in his equation much more than what he had originally designed it for: he read the existence of antimatter, again a conceptual and practical revolution. In this sense, the equation is the perfect example of the power of mind over matter — every fan of Buddhism should take a five-year university course to understand this equation.

In 1928, Quantum Mechanics and Relativity had both been well formulated and explored. But attempts to reconcile them had been awkward and fruitless. To describe a quantum and relativistic particle, there existed an equation, the Klein-Gordon equation:

$$(\square + m^2)\phi = 0$$

$(i\gamma^{\mu}\partial_{\mu} - m)\psi = 0$

However, it remained purely theoretical, because it does not apply to known particles like the electron or the proton, which make up the atoms. In any case, you will agree with me that it is not a graphical success, with its clumsy $\Box$ and ϕ.

The reason why this equation does not correctly describe the electron, for example, is related to a property of the particles: their spin. As the word evokes, the frequently used image to represent the spin is that of a particle spinning like a top. In reality, the spin is a purely quantum quantity, which does not require the idea of a solid top: an elementary particle, which is by definition point like, can have a spin, and this is the case for the electron.

In particular, the spin impacts the way a particle interacts with a magnetic field: it is possible to determine the spin of an electron by measuring its deviation by a certain configuration of magnetic field. Experimentally the measurement always returns one of two values (and not a continuum of values), which are the associated to two states of spin, noted up (↑), or down (↓).

Back in 1928, the spin had been known for a few years and it was well explained by Quantum Mechanics. However, the Klein-Gordon equation ignored the spin: it seemed impossible to describe an electron at the same time as quantum and relativistic, although these two theories have already revolutionized physics!

The equation first just fulfills its role

In a stroke of genius, Dirac realized that the Klein-Gordon equation is maybe too simple, and perhaps hides the subtlety of the spin. He imagined that the operation $(\Box + m^2)$ might be decomposed into two factors $(D) \times (D)$, where each factor (D) would be sensitive to the spin but their product would not.

A quick analogy: take the number "4". I could have obtained it by taking the square of the number 2: $4 = (2) \times (2)$. But I could also have taken the square of the number –2: $4 = (-2) \times (-2)$. When taking the square, I lost the information of the sign between + 2 and – 2. Similarly, the Klein-Gordon equation would lose the information on the spin.

Dirac understood that he could indeed perform such a decomposition and write the factor (D) if he treated the (↑) and (↓) components of the electron separately, if only he combined them harmoniously with the help of the coefficients γ_μ which you see in the equation. In passing, he discovered the amusing mathematical properties of these coefficients for which $A \times B$ are not equal to $B \times A$...

This is not an attempt to give you a complete explanation, but to make you feel that already at this stage, inventing this equation marks a deep culture, a sharp imagination, and a wide freedom of thought.

The resulting equation describes the electron and allows for immediate applications, for example the fine calculation of the energy levels in atoms, in perfect agreement with the measurements.

Quite physical solutions

The story could have ended there, Dirac would have certainly received his Nobel Prize, and his equation would have been taught in universities. But the best was yet to come.

Indeed, he noticed that his equation possessed two types of solutions: the first type corresponded to the description of electrons, as he wanted in the first place. The second type was a mathematical function which was a solution of the equation, but with parameters such that it was impossible to interpret it in terms of electrons. The energy of such particles would in fact be negative...

You have to realize that this is a rather common situation in physics. To model some phenomenon, a physicist has struggled to build a nice equation. If this step is successful, the quantity or the function which supports the model is a solution of the equation, by construction. But there may exist other solutions, which do not correspond to anything physical. These are spurious solutions, which do not spoil the initial merit of the equation. One only needs to accompany the equation instruction manual with a warning: "in the subsequent calculation, be careful to exclude the spurious, non-physical solutions.".

This is what any average physicist would have done, or even every very good physicist who would have invented this equation. But Dirac was not satisfied with this. What does this extra solution mean? What could be the physical meaning of a negative energy electron? He realized that the extra solution might as well describe a real particle with the same mass as the electron, but with the opposite electric charge.

Dirac believed in this. These new solutions are not spurious. Everything is too consistent to be useless: these unexpected solutions must describe real particles, which do exist in nature. Dirac invented antiparticles: the positron would become the electron antiparticle, the antiproton that of the proton. A positron and an antiproton could combine to form an anti-atom, which in turn could form anti-molecules…

Antimatter exists!

With this founding move, Dirac invented antimatter, a world which mirrors ordinary matter. He also realized that it is possible to create particle-antiparticle pairs out of pure energy, and conversely a particle and its antiparticle can annihilate into energy.

When Carl Anderson, in 1932, observed in the fragments of cosmic rays the track of an "electron", but with the opposite charge, he knew that he had just discovered the positron predicted by Dirac experimentally. Since, we have observed the antiparticles of all known particles. Particles and antiparticles are studied on an equal footing in physics experiments. The positron has found applications for medical imaging — PET scan, or Positron Emission Tomography — and material control.

From the theoretical point of view, the existence of antiparticles marks a revolution. Since a particle-antiparticle pair can be created from pure energy, this means that events which imply enough energy concentrated in a very small space are able to create such pairs. Let us consider the collision of two particles A and B: if the collision energy is large enough (and a few other conditions), it may lead to the creation of an electron-positron pair, along the process:

$$A + B \rightarrow A + B + e^{+} + e^{-}$$

Consequently, the reactions among elementary particles do not conserve the number of particles! This is very different from the "classical" intuition where particles would be billiard balls. Instead, it is a rather strange game where the number of balls varies after each impact!

Believing in an equation

This line of reasoning extends to any known particle, or even to any *unknown* one: with enough energy, it is possible to create it as a particle-antiparticle pair. For completeness, let me add that some particles like the photon (neutral for the electric charge and other charges) are their own antiparticles and can therefore be single-produced in such a reaction.

The collisions between known particles have thus become a way of searching for unknown physics: it is the principle behind the experiments conducted at large accelerators like the LHC at CERN, which collides protons at a very high energy (per particle). Contrary to what a classical intuition could make believe, the outgoing particles which are observed after a proton-proton collision are not fragments of the initial protons. They are essentially particles created from the energy of the collision.

How can a single equation extend in one move the field of knowledge, much beyond what it was invented for? Is it because it contains in itself a large part of truth? Or does it only reveal facts that we already knew, however without interpreting them?

But above all, how could Dirac have such a faith in his equation? Why did his (immense) culture and his practice bring him to avoid reacting like the average physicist, and to disentangle the circumstantial and the fundamental?

Elegant in its graphical appearance and deep in its message, condensed in its language and creative in its design, Dirac's equation moves me like the most beautiful poetry, nothing better describes "… such stuff as dreams are made on…".

Chapter 14
Feynman diagrams

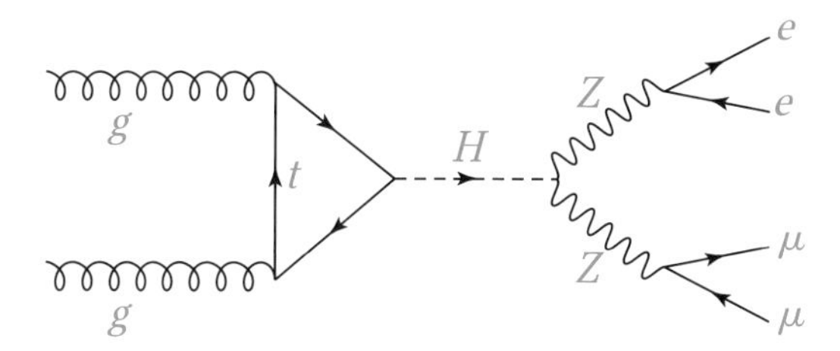

This chapter is not about one particular equation, like the previous ones, but rather a system of small sketches which are used to *represent* equations, almost all the equations which govern the motions and interactions of elementary particles. These are Feynman diagrams, named after their inventor Richard Feynman (1918–1988). They constitute a language which can represent, or even encode, some otherwise extremely complex equations. Although they do not modify the underlying theory, these diagrams have eased the calculations, then the reasoning itself, so much that they dominate the whole practice of Particle Physics. This is an excellent opportunity for asking the question of what comes first: a theory or its representation? What role do our mental images play, even the most abstract ones, in our vision of nature?

The framework of all this modern Particle Physics is Quantum Field Theory (QFT), meaning relativistic fields. It was born from the more or less harmonious union of special relativity (that of $E = mc^2$) and Quantum Mechanics. After the spectacular step forward of Dirac's equation, the quantum relativistic theory developed rapidly, and the very notion of particle, interactions, and even the vacuum, underwent a radical change. One of the most striking consequences is that the number of particles is not conserved through interactions. Quantum Mechanics had already challenged all intuitive notions,

e
e

like position, velocity, energy, frequency, but a particle remained a particle. After Dirac and the discovery of antiparticles, nothing prevents that during an interaction, energy transforms into a particle-antiparticle pair.

In the 1930's, scientists cast all this in equations, allowing to calculate reaction probabilities which took these effects into account. However, these calculations were long and cumbersome. The variables used were mathematical operators which obeyed a complex an inelegant grammar. Only skilled physicists managed to formulate theoretical predictions for simple interactions between electrons, photons, and atomic nuclei, which were spectacularly confirmed by experiments: the conversion of high energy photons into electron-positron pairs, the radiation of photons by electrons passing in the vicinity of a nucleus, etc.

Tedious calculations...

Encouraged by these first successes, physicists undertook more ambitious calculations, but they soon realized that for some interaction probabilities, the result of the calculation tends to become... infinite. The problem is that the *quantum* character of the theory allows for energy fluctuations for example, and the *relativistic* character allows the conversion of energy into matter. Here and there, real or "virtual" particles pop-up, whose number and energy cannot be controlled, and the calculations lose all their meaning. It would take years of trial and error (from the end of the 1930 to about 1960) to build "renormalization", a rather suspect procedure, mathematically speaking, but clear enough on the physical side to restore the efficiency of relativistic quantum theory.

In this way, calculations have become extremely tedious, the equations are long chains of mathematical operators, and for the calculation of a single interaction probability, the mere bookkeeping of equations and their terms becomes the dominant limiting factor.

A little diagram for a long calculation

Here enters Richard Feynman. A fantastic physicist, he made several fundamental contributions. In the early 1950's, he was a bright young man, who just participated in the Manhattan project, the making of the nuclear weapon at Los Alamos. In particular, the design of the bomb required, for the first time, numerical computations on the very first digital computers. These computers were very limited, very slow, their programming was all but basic, and Feynman used all his ingenuity identifying repetitions, dependencies and interdependencies, and organized these computations in the optimal way.

Back to fundamental physics after the war, he soon got tired of the tedious aspect of relativistic quantum calculations, and he sought a structure, some kind of grammar. He soon noticed that each term corresponded to a simple classical image, for instance:

e ——→—— e $\qquad$ γ ∿∿∿∿∿ γ

for the terms of propagation of an electron (left), and a photon (right), and:

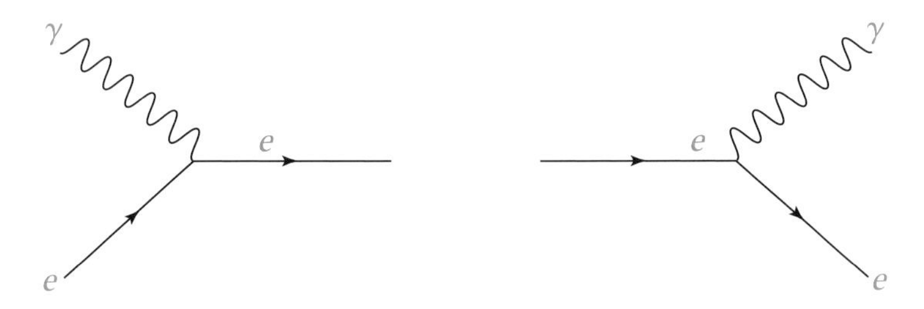

for the absorption or the emission of a photon by an electron.

He formulated the implicit rules of the calculations, and the corresponding rules for the little diagrams.

Finally, he translated into diagrams the calculation already known by that time. For example, the probability for the diffusion of a photon on an electron is obtained from the following formula, already a rather complex one:

$$\bar{u}(p') \left[\epsilon'^{*}_{\nu}\, (ie\gamma^{\nu})\, \frac{i(\not{p} + \not{k} + m)}{(p+k)^2 - m^2}\, (ie\gamma^{\mu})\, \epsilon_{\mu} \right] u(p)$$

which can be represented by the very simple diagram:

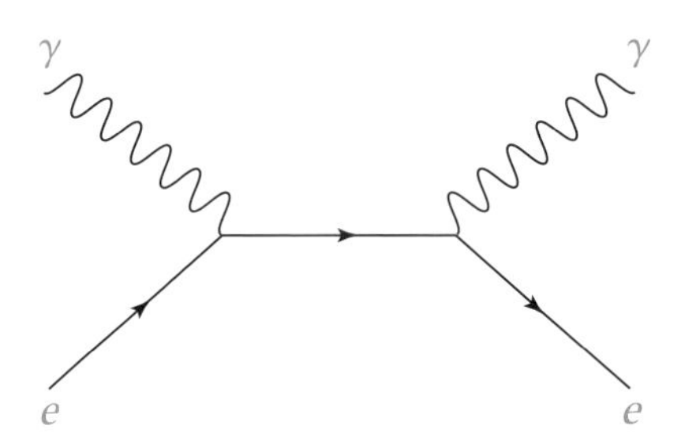

At this stage, Feynman diagrams were yet a description utility. Then they became a reminder, since it is much easier to count diagrams, which are visually different, than equations. In the end, they became a real language. This is the ideal tool: the diagrams offered a true translation of complex calculations into a simple and intuitive visual form.

A powerful language

The success of Feynman diagrams was immediate. Renormalization, the procedure which allows to give a meaning to calculations whose results would otherwise be infinite, was easily illustrated in terms of diagrams, to the extent that one could wonder which is the more fundamental, the theory or the diagrams? Feynman himself brought essential contributions to the implementation of renormalization.

Since that time, it became impossible to enter the office of a particle physicist, theorist or experimentalist, without finding a few Feynman diagrams on the blackboard.

Elaborated in the first place for electrodynamics (i.e. the electromagnetic interactions between electrons, positrons, photons, nuclei…), quantum field theory soon included all interactions (strong nuclear, weak nuclear) and all elementary particles. Each time the diagrams proved easy to adapt, still representing complex processes in an intuitive way, hiding calculations whose development would be often extremely long and for sure not very interesting except for their result.

For instance, the production and disintegration of a Higgs boson at the LHC accelerator at CERN are computed from the diagram at the beginning of this chapter.

Thanks to its precise grammar, the description by diagrams is also incredibly helpful for numerical calculations. Let us imagine that a theorist has invented a new theory, from fundamental principles, and we would like to compare the predictions of this theory with the observation of the collisions produced by a particle accelerator, say the CERN LHC. If this theory is predictive, it indicates the elementary particles and their interaction laws. To allow the calculation of experimental quantities, the theorist formulates the "Feynman rules" of the theory. Other physicists then translate these rules into computer code, and add them to large programs which can automatically generate all the Feynman diagrams necessary for calculations. Finally, experimentalists at the LHC use the output of the previous code to generate simulated events, in order to compare them to the real events observed in particle collisions. Today, this suite of programs has reached amazing performances, reproducing in simulations all the known physics including its most subtle effects.

Dreaming of Feynman diagrams

There are other cases in science where diagrams represent calculation or structures, like Venn diagrams in set theory, Young and Dynkin diagrams in group theory, and also outside pure science, like in juggling where the crossings of arms are noted in diagrams... But I don't think that any of these systems has acquired the same status as a language than Feynman diagrams in particle physics.

Structuralist philosophers, Claude Lévy-Strauss to begin with, have also ventured to describe human interactions with diagrams, without much success, at least to my taste. Jacques Lacan, of course, always craving for scientific acknowledgment, has called diagrams to help, but it is difficult to see there a precise, non-ambiguous description of an effect or a concept. I challenge an analyst who would not have read (and memorized!) the *Ecrits*, to tell me without warning the meaning of this or that diagram. But for a physicist, a Feynman diagram is always the representation of a precise, unique equation.

We think, we talk, and above all we imagine with Feynman diagrams. I even happened to dream with Feynman diagrams: during a somehow complicated period of my personal life, and at the same time a moment of intense professional activity. This was a naïve dream, where I tried to describe with diagrams some difficult personal choices. In the dream, I was indeed aware that these problems were difficult to "renormalize", that multiple choices comprised a large number of diagrams, infinitely many maybe. Alas, the methods of field theory are of little help for human matters, and after awakening, only the rather unpleasant feeling of a sterile mix between dream and reality remained, between private and professional worlds...

Fortunately, these dreams were very rare, even surprisingly rare, for me at least.

Chapter 15

The standard model

$$L = -\frac{1}{4} F_{\mu\nu} F^{\mu\nu} + i\overline{\psi} \not{D} \psi + h.c. + \psi_i \gamma_{ij} \psi_j \varphi$$

$$+ h.c. + \left| D_\mu \varphi \right|^2 - \left[\frac{1}{2} \mu^2 \left| \varphi \right|^2 + \frac{1}{4} \lambda \left| \varphi \right|^4 \right]$$

This formula is the basis of the "Standard Model" of Particle Physics, the best theory we have today to describe the elementary bricks of nature, and the structure of their interactions. In short, all that allows them to interact, assemble and construct the Universe we know, from the Big-Bang down to today.

The equation is famous enough that it is printed on the souvenir T-shirts and mugs sold at CERN, the large particle physics center near Geneva, straddling the French-Swiss border.

In the CERN cafeteria

I first came to CERN in 1979, at the very beginning of my PhD, when I accompanied my supervisor, Marcel Banner. At the CEA-Saclay[1], our team had built a prototype detector which we wanted to test in a beam of particles. We arrived at CERN on a Monday morning, and we immediately went to the experimental hall where the beam was planned. I only knew

[1] The French CEA ("Commissariat à l'Energie Atomique") is the equivalent of the US DOE for nuclear energy and research. CEA-Saclay is a large research center near Paris which includes my home laboratory.

$$L = -\frac{1}{4}F_{\mu\nu}F^{\mu\nu} + i\bar{\psi}\not{D}\psi + h.c. + \psi_i\gamma_{ij}\psi_j\varphi + h.c. + |D_\mu\varphi|^2 - \left[\frac{1}{2}\mu^2|\varphi|^2 + \frac{1}{4}\lambda|\varphi|^4\right]$$

CERN through texts and photographs, and I had barely enough time to glimpse the reality of this hall, long ceiling-less corridors between gray concrete blocks, with at the end complicated apparatus swarming with cables and pipes.

In the beamline reserved for us, there was almost nothing: we could only notice the pipe from which the particles would come, and beyond that, nothing but the floor and two walls, all made of raw concrete. Marcel asked me if I could handle a drill, and I said yes. He then put into my hands an enormous hammer drill for concrete. A test beam indeed requires a certain number of detectors which record the arrival time and position of the beam particles. These "counters" are fixed to the floor and the walls all along the beamline, between the delivery pipe, which comes from the accelerator itself, to the zone where the apparatus on test sits.

Hence I started my first stay at CERN… drilling holes with a hammer-drill into a rather hard concrete! Marcel was a workaholic, and everything moved forward without a minute of rest: hanging the counters, laying cables over tens of meters towards the small control room where the computer which would record the test data was, adjusting all the electronics, trimming the amplifiers, the logical circuits, testing the data acquisition software… Arrived on Monday morning, we completed the installation around Wednesday noon, only pausing for quick meals at the nearby cafeteria. Everything was operational, ready to receive the first beam particles and test our prototype.

I was in a strange state, arms aching, body still vibrating from the shaking of the drill, head full of all I had just learnt by seeing my supervisor swiftly tuning complex electronics. Before going to bed, we went for a beer at the main cafeteria, and I collapsed in an armchair in front of a coffee table. Around us, people spoke English, of course, but all other kinds of languages as well. Coffee in hand, people were talking, debating, scribbling on paper napkins. A couple of tables away, Murray Gell-Mann (1969 Nobel Prize) was chatting with Carlo Rubbia (already rather famous, he had not yet received his Nobel Prize, which he got in 1984). In this strange state of fuzziness and lucidity induced by the lack of sleep, I thought: This is *the* place.

The Lagrangian returns

Let us get back to the equation of the Standard Model. The form "$L = \ldots$" indicates that this is a Lagrangian, this operator we have already encountered in Chapter 2. The right term then contains the mathematical description of the constituents of matter and their interactions, for this theory. The first two terms correspond, roughly speaking, to electromagnetism after the invention of Dirac's equation, the first one for photons, the second one for electrons.

In the next terms, you will note the letter "ϕ", which denotes the "Higgs field", better known as the "Higgs boson". This new particle, essential for the coherence of the Standard Model, appeared first in 1964 in a simpler context, then after a few years the equation took the form above.

Between 1930 and 1960, nuclear physics had made much progress. Experiments studying cosmic rays — these particles which come from the sky — then experiments at particle accelerators, discovered many new particles, which are not constituents of ordinary matter, and new interactions.

Theory also evolved, or rather theories, since in the beginning each type of interaction required its own theory: electromagnetism described electrons and photons, weak interactions described the radioactive decays, and strong interactions described the cohesion of atomic nuclei. Around 1960, physicists could understand that there were similarities among these interactions, but they did not manage to provide them with a truly consistent framework. In particular, the theories of electromagnetism and of weak interactions appeared very close, but separated by an insurmountable obstacle.

Two articles appeared in 1964 that brought the solution to the problem, the first by two Belgian theorists, Robert Brout and François Englert, the second by the Scottish theorist Peter Higgs. The key point is to introduce a new "field", with strange properties: its mere existence modifies the properties of all other particles, including their mass and their interactions. This new theory bypassed in a most elegant way the wall that separated electromagnetism from weak interactions: it is indeed a single interaction, but the mere presence of the Higgs field makes them appear under two distinct forms. This "disguise" takes place softly, and in a consistent way over all particles.

Most importantly, the theory formulated a series of unexpected predictions, which could be tested at particle accelerators, existing or in the future. There would exist a particle called "Z^0", similar to the photon but with a very high mass (for the time), around a hundred times the mass of a proton. In addition, the new field could show up as another new particle, which would very quickly be known in the community as the "Higgs boson".

From this starting point, the main obstacle was removed, and the Standard Model was progressively built, following the evolution of particle accelerators: some already known particles were interpreted in this framework (muons, neutrinos), and other were then discovered (quarks which make up protons and neutrons, the other types of quarks), which each time nicely found their place in the model framework. In 1984, the Z^0 was discovered at CERN, and in 1995 the last and most massive quark of the model was identified at the Fermi laboratory close to Chicago.

Hunting the Higgs

But where was the *Higgs*? The Standard Model theory described all its properties… except the most important one from the experimental point of view: its mass. It barely indicated a range of possible masses, with a highest value far beyond the reach of the accelerators of the 1970's. Hence each time a new accelerator came into operation, physicists hunted the *Higgs*, but somehow haphazardly, since they did not know if the energy involved would be enough to create it. In 1989 (14th of July, an easy-to-remember date for the French!), CERN launched the operation of the LEP, a very large electron-positron "collider", 8.6 km diameter. Until 2000, the LEP achieved an impressive number of measurements which completed and refined the Standard Model, but the *Higgs* remained out of reach.

Already in 1984, a few tens of physicists held a workshop in Lausanne to discuss the post-LEP era: why not reuse the same tunnel, and replace the LEP by a proton-proton collider, which could reach much higher energies, the LHC? If the LEP would discover the *Higgs*, the LHC would allow to better explore its properties at the core of the Standard Model. If not, the LHC should have enough energy to cover the whole possible mass range, and it would almost surely discover the *Higgs*, if it really exists.

I attended this Lausanne workshop, I was a PhD student and this was the year of my thesis defense. The atmosphere was quite enthusiastic and creative, but I cannot say we really believed that a LHC would really come true…

We were just becoming aware of the huge challenges raised by the design and realization of such a project. Both on the accelerator side, and on the side of the experiments which would be built to observe these collisions, the performances should be much higher than the know-how of the time. From the field in the magnets, to the accuracy of the detectors, to the speed of the electronics, and the amount of data to process, etc.: we attempted back-of-the-envelope estimates of all the necessary specifications, and each time the result looked rather scary. But taking a closer look, and taking into account the predictable technological progresses, why not?

A bright confirmation

The rest is well known: a decade of technical design and diplomacy to convince CERN members states, and several other countries, then another decade of construction, at CERN and in a large number of laboratories in the entire world, and in 2008 the LHC received its first beams. After a somehow hectic startup, the accelerator and the experiments reached a good level of operation in 2011 and 2012, and the teams could actively seek the *Higgs* in the abundantly produced collisions.

The 4th of July 2012 (an easy-to-remember date for the US!), after more than 20 years of preparation work, and only 2 years of LHC operation, the CERN directors called for an exceptional seminar in the large auditorium with the title “Search of the Higgs boson at LHC by the ATLAS and CMS experiments”. ATLAS and CMS are the two large collaborations (about 3000 physicists each) which exploit two huge detectors located at two opposite points along the LHC ring, where the proton collisions are produced. The collaborations are independent, and are even competitors, and their preliminary results are kept secret.

Like all my ATLAS colleagues, I knew the result that Fabiola Gianotti, the ATLAS experiment spokesperson, would present, but I didn’t know what Joe Incandela, the CMS spokesperson, would say. Only the CERN director general and the two spokespersons knew both results. In ATLAS, the data analysis had shown evidence for a new particle which did look like the Higgs, with a high confidence level. We could feel that if such a seminar was organized, inviting the press and the member state delegates, CMS must also have a positive result. And when we entered the auditorium, Peter Higgs and François Englert were there, invited by CERN[2]. But would the two results be compatible? For us, specialists, what would be, in detail, the evidence seen by the other collaboration, if any? We were ready to compare each graphic plot, each number, to what we noted in our experiment.

Joe started presenting CMS results, for one hour, then in turn Fabiola presented for ATLAS. A long round of applause filled the auditorium, then outside we rushed to the press conference, and after it ended each of us was asked for comments by journalists of her/his country, during hours.

Later in the afternoon, I was alone, in a quiet place. I was still in a state of confusion and extreme lucidity due to lack of sleep, after days and nights spent analyzing the data. And I thought of my first visit to CERN, and of the holes in the concrete blocks. There we were in 2012, and the *Higgs*, invented by theorists in 1964, was now experimentally confirmed. Its mass is 125 GeV, about 130 times the mass of a proton. I could now go for sleep.

---

[2] Robert Brout passed away in 2011.

Epilogue

The limits of the standard model... and future physics

$$[?]\Psi = 0$$

The Standard Model has been completed by the experimental confirmation of the existence of the Higgs boson. Peter Higgs and François Englert received the Nobel Prize in 1993. But the quest does not stop there.

In fact, I have left one point untold: the equation shown in the previous chapter is only a template equation for the Standard Model. In practice, the model contains many arbitrary parameters and a few casual arrangements, and the complete equation rather looks like that reproduced on the next page[1]!

Above, this is the equation all physicists can dream of: perfectly streamlined, with zero free parameters, the "Truth of the World" revealed, in a sense. As you can see, we are still quite far from it, with our Standard Model!

Standard, really?

Before outlining its limits, I would like to come back to the expression "Standard Model". The expression is interesting, because it speaks of a globalized science, in every sense of the word. Globalized, since all the physicists in the world use the same formalism, the same notations. They have agreed to say that a certain theory is common to all, and that it accounts for all experimental measurements performed up to now. One could see here a form of imperialism, where any "non-standard" idea would be rejected. Indeed, like many of my colleagues, I happen to receive letters or messages from persons who think they have discovered the hidden sense of the world, and

[1] From Gutierrez : http://nuclear.ucdavis.edu/~tgutierr/files/sml.pdf

$$-\tfrac{1}{2}\partial_\nu g^a_\mu \partial_\nu g^a_\mu - g_s f^{abc}\partial_\mu g^a_\nu g^b_\mu g^c_\nu - \tfrac{1}{4}g_s^2 f^{abc} f^{ade} g^b_\mu g^c_\nu g^d_\mu g^e_\nu +$$
$$\tfrac{1}{2}ig_s^2(\bar{q}^\sigma_i \gamma^\mu q^\sigma_j) g^a_\mu + \bar{G}^a\partial^2 G^a + g_s f^{abc}\partial_\mu \bar{G}^a G^b g^c_\mu - \partial_\nu W^+_\mu \partial_\nu W^-_\mu -$$
$$M^2 W^+_\mu W^-_\mu - \tfrac{1}{2}\partial_\nu Z^0_\mu \partial_\nu Z^0_\mu - \tfrac{1}{2c_w^2}M^2 Z^0_\mu Z^0_\mu - \tfrac{1}{2}\partial_\mu A_\nu \partial_\mu A_\nu - \tfrac{1}{2}\partial_\mu H \partial_\mu H -$$
$$\tfrac{1}{2}m_h^2 H^2 - \partial_\mu \phi^+ \partial_\mu \phi^- - M^2 \phi^+ \phi^- - \tfrac{1}{2}\partial_\mu \phi^0 \partial_\mu \phi^0 - \tfrac{1}{2c_w^2} M \phi^0 \phi^0 - \beta_h[\tfrac{2M^2}{g^2} +$$
$$\tfrac{2M}{g}H + \tfrac{1}{2}(H^2 + \phi^0\phi^0 + 2\phi^+\phi^-)] + \tfrac{2M^4}{g^2}\alpha_h - igc_w[\partial_\nu Z^0_\mu (W^+_\mu W^-_\nu -$$
$$W^+_\nu W^-_\mu) - Z^0_\nu (W^+_\mu \partial_\nu W^-_\mu - W^-_\mu \partial_\nu W^+_\mu) + Z^0_\mu (W^+_\nu \partial_\nu W^-_\mu -$$
$$W^-_\nu \partial_\nu W^+_\mu)] - igs_w[\partial_\nu A_\mu (W^+_\mu W^-_\nu - W^+_\nu W^-_\mu) - A_\nu (W^+_\mu \partial_\nu W^-_\mu -$$
$$W^-_\mu \partial_\nu W^+_\mu) + A_\mu (W^+_\nu \partial_\nu W^-_\mu - W^-_\nu \partial_\nu W^+_\mu)] - \tfrac{1}{2}g^2 W^+_\mu W^-_\mu W^+_\nu W^-_\nu +$$
$$\tfrac{1}{2}g^2 W^+_\mu W^-_\nu W^+_\mu W^-_\nu + g^2 c_w^2 (Z^0_\mu W^+_\mu Z^0_\nu W^-_\nu - Z^0_\mu Z^0_\mu W^+_\nu W^-_\nu) +$$
$$g^2 s_w^2 (A_\mu W^+_\mu A_\nu W^-_\nu - A_\mu A_\mu W^+_\nu W^-_\nu) + g^2 s_w c_w [A_\mu Z^0_\nu (W^+_\mu W^-_\nu -$$
$$W^+_\nu W^-_\mu) - 2A_\mu Z^0_\mu W^+_\nu W^-_\nu] - g\alpha[H^3 + H\phi^0\phi^0 + 2H\phi^+\phi^-] -$$
$$\tfrac{1}{8}g^2\alpha_h[H^4 + (\phi^0)^4 + 4(\phi^+\phi^-)^2 + 4(\phi^0)^2\phi^+\phi^- + 4H^2\phi^+\phi^- + 2(\phi^0)^2H^2] -$$
$$gMW^+_\mu W^-_\mu H - \tfrac{1}{2}g\tfrac{M}{c_w^2}Z^0_\mu Z^0_\mu H - \tfrac{1}{2}ig[W^+_\mu(\phi^0\partial_\mu\phi^- - \phi^-\partial_\mu\phi^0) -$$
$$W^-_\mu(\phi^0\partial_\mu\phi^+ - \phi^+\partial_\mu\phi^0)] + \tfrac{1}{2}g[W^+_\mu(H\partial_\mu\phi^- - \phi^-\partial_\mu H) - W^-_\mu(H\partial_\mu\phi^+ -$$
$$\phi^+\partial_\mu H)] + \tfrac{1}{2}g\tfrac{1}{c_w}(Z^0_\mu(H\partial_\mu\phi^0 - \phi^0\partial_\mu H) - ig\tfrac{s_w^2}{c_w}MZ^0_\mu(W^+_\mu\phi^- - W^-_\mu\phi^+) +$$
$$igs_w MA_\mu(W^+_\mu\phi^- - W^-_\mu\phi^+) - ig\tfrac{1-2c_w^2}{2c_w}Z^0_\mu(\phi^+\partial_\mu\phi^- - \phi^-\partial_\mu\phi^+) +$$
$$igs_w A_\mu(\phi^+\partial_\mu\phi^- - \phi^-\partial_\mu\phi^+) - \tfrac{1}{4}g^2 W^+_\mu W^-_\mu[H^2 + (\phi^0)^2 + 2\phi^+\phi^-] -$$
$$\tfrac{1}{4}g^2\tfrac{1}{c_w^2}Z^0_\mu Z^0_\mu[H^2 + (\phi^0)^2 + 2(2s_w^2 - 1)^2\phi^+\phi^-] - \tfrac{1}{2}g^2\tfrac{s_w^2}{c_w}Z^0_\mu\phi^0(W^+_\mu\phi^- +$$
$$W^-_\mu\phi^+) - \tfrac{1}{2}ig^2\tfrac{s_w^2}{c_w}Z^0_\mu H(W^+_\mu\phi^- - W^-_\mu\phi^+) + \tfrac{1}{2}g^2 s_w A_\mu\phi^0(W^+_\mu\phi^- +$$
$$W^-_\mu\phi^+) + \tfrac{1}{2}ig^2 s_w A_\mu H(W^+_\mu\phi^- - W^-_\mu\phi^+) - g^2\tfrac{s_w}{c_w}(2c_w^2 - 1)Z^0_\mu A_\mu\phi^+\phi^- -$$
$$g^1 s_w^2 A_\mu A_\mu\phi^+\phi^- - \bar{e}^\lambda(\gamma\partial + m_e^\lambda)e^\lambda - \bar{\nu}^\lambda\gamma\partial\nu^\lambda - \bar{u}^\lambda_j(\gamma\partial + m_u^\lambda)u^\lambda_j -$$
$$\bar{d}^\lambda_j(\gamma\partial + m_d^\lambda)d^\lambda_j + igs_w A_\mu[-(\bar{e}^\lambda\gamma^\mu e^\lambda) + \tfrac{2}{3}(\bar{u}^\lambda_j\gamma^\mu u^\lambda_j) - \tfrac{1}{3}(\bar{d}^\lambda_j\gamma^\mu d^\lambda_j)] +$$
$$\tfrac{ig}{4c_w}Z^0_\mu[(\bar{\nu}^\lambda\gamma^\mu(1 + \gamma^5)\nu^\lambda) + (\bar{e}^\lambda\gamma^\mu(4s_w^2 - 1 - \gamma^5)e^\lambda) + (\bar{u}^\lambda_j\gamma^\mu(\tfrac{4}{3}s_w^2 -$$
$$1 - \gamma^5)u^\lambda_j) + (\bar{d}^\lambda_j\gamma^\mu(1 - \tfrac{8}{3}s_w^2 - \gamma^5)d^\lambda_j)] + \tfrac{ig}{2\sqrt{2}}W^+_\mu[(\bar{\nu}^\lambda\gamma^\mu(1 + \gamma^5)e^\lambda) +$$
$$(\bar{u}^\lambda_j\gamma^\mu(1 + \gamma^5)C_{\lambda\kappa}d^\kappa_j)] + \tfrac{ig}{2\sqrt{2}}W^-_\mu[(\bar{e}^\lambda\gamma^\mu(1 + \gamma^5)\nu^\lambda) + (\bar{d}^\kappa_j C^\dagger_{\lambda\kappa}\gamma^\mu(1 +$$
$$\gamma^5)u^\lambda_j)] + \tfrac{ig}{2\sqrt{2}}\tfrac{m_e^\lambda}{M}[-\phi^+(\bar{\nu}^\lambda(1 - \gamma^5)e^\lambda) + \phi^-(\bar{e}^\lambda(1 + \gamma^5)\nu^\lambda)] -$$
$$\tfrac{g}{2}\tfrac{m_e^\lambda}{M}[H(\bar{e}^\lambda e^\lambda) + i\phi^0(\bar{e}^\lambda\gamma^5 e^\lambda)] + \tfrac{ig}{2M\sqrt{2}}\phi^+[-m_d^\kappa(\bar{u}^\lambda_j C_{\lambda\kappa}(1 - \gamma^5)d^\kappa_j) +$$
$$m_u^\lambda(\bar{u}^\lambda_j C_{\lambda\kappa}(1 + \gamma^5)d^\kappa_j] + \tfrac{ig}{2M\sqrt{2}}\phi^-[m_d^\lambda(\bar{d}^\lambda_j C^\dagger_{\lambda\kappa}(1 + \gamma^5)u^\kappa_j) - m_u^\kappa(\bar{d}^\lambda_j C^\dagger_{\lambda\kappa}(1 -$$
$$\gamma^5)u^\kappa_j] - \tfrac{g}{2}\tfrac{m_u^\lambda}{M}H(\bar{u}^\lambda_j u^\lambda_j) - \tfrac{g}{2}\tfrac{m_d^\lambda}{M}H(\bar{d}^\lambda_j d^\lambda_j) + \tfrac{ig}{2}\tfrac{m_u^\lambda}{M}\phi^0(\bar{u}^\lambda_j\gamma^5 u^\lambda_j) -$$
$$\tfrac{ig}{2}\tfrac{m_d^\lambda}{M}\phi^0(\bar{d}^\lambda_j\gamma^5 d^\lambda_j) + \bar{X}^+(\partial^2 - M^2)X^+ + \bar{X}^-(\partial^2 - M^2)X^- + \bar{X}^0(\partial^2 -$$
$$\tfrac{M^2}{c_w^2})X^0 + \bar{Y}\partial^2 Y + igc_w W^+_\mu(\partial_\mu\bar{X}^0 X^- - \partial_\mu\bar{X}^+ X^0) + igs_w W^+_\mu(\partial_\mu\bar{Y}X^- -$$
$$\partial_\mu\bar{X}^+ Y) + igc_w W^-_\mu(\partial_\mu\bar{X}^- X^0 - \partial_\mu\bar{X}^0 X^+) + igs_w W^-_\mu(\partial_\mu\bar{X}^- Y -$$
$$\partial_\mu\bar{Y}X^+) + igc_w Z^0_\mu(\partial_\mu\bar{X}^+ X^+ - \partial_\mu\bar{X}^- X^-) + igs_w A_\mu(\partial_\mu\bar{X}^+ X^+ -$$
$$\partial_\mu\bar{X}^- X^-) - \tfrac{1}{2}gM[\bar{X}^+ X^+ H + \bar{X}^- X^- H + \tfrac{1}{c_w^2}\bar{X}^0 X^0 H] +$$
$$\tfrac{1-2c_w^2}{2c_w}igM[\bar{X}^+ X^0\phi^+ - \bar{X}^- X^0\phi^-] + \tfrac{1}{2c_w}igM[\bar{X}^0 X^-\phi^+ - \bar{X}^0 X^+\phi^-] +$$
$$igMs_w[\bar{X}^0 X^-\phi^+ - \bar{X}^0 X^+\phi^-] + \tfrac{1}{2}igM[\bar{X}^+ X^+\phi^0 - \bar{X}^- X^-\phi^0]$$

who often claim that they are ignored by the establishment, prevented from publishing, victims of the intellectual dictatorship of the scientific community. In their imagination, the "Standard Model" is nothing but a dogma.

Yet it is just the opposite! It is, again, a form of humility and open-mindedness. We agree on what is *standard* to better discuss *what could be non-standard.* The model is constantly analyzed, scrutinized, dissected, in its smallest details. Have we explored all the predictions, at least all the *testable* ones? Could we imagine a new experiment that could test it in some other way?

The large consistency of the model allows to test it from very diverse points of view: in the measurements of proton collisions at very high energy at the LHC, but also in high precision measurements on a few atoms, or again in the observation of the cosmos. Wherever it comes from, as soon as a new measurement is made, the community immediately confronts it with the Standard Model. Does it agree with the prediction of the model for this effect, or not? Contrary to the idea of those who denounce a potential scientific imperialism, the dream of every physicist is not to confirm the official theory, but on the contrary to find in a new measurement a definite sign that the Standard Model is faulty.

Furthermore, even if we use it every day, we are well aware of its limits and of its inconsistencies. With the discovery of the Higgs boson in 2012 at the CERN LHC, the chapter of the "elementary particles" looks complete and consistent, but its main ingredients remain unexplained. For example, matter comprises twelve elementary particles, belonging to three families of four particles each: the first family is that of which ordinary matter is made, the other two are replicas of the first, containing particles with similar properties but with higher masses. Why are there these three families? Nobody knows. We can just see that the model describes the results of experiments well; in particular, the three families imply some subtle symmetry properties which could not be present with only one or two families, and which have been seen experimentally.

The particle masses, interpreted as the intensity of their interaction with the Higgs field, are free parameters of the theory, along with all the other interaction intensities, and we can only measure them experimentally. Thus we count twenty-seven free parameters to fix. Similarly, the mathematical structures which drive the interactions are well identified, but why these rather than different ones?

If we accept this part of arbitrariness, of experimental fixing of the parameters, we have in hand a coherent and powerful theory, which has been tested by tens of diverse measurements and never been proved wrong, at least at the small scales of distance, in the world of elementary particles.

But where is gravitation?

Another shortcoming: this quantum theory does not include gravitation. Gravitation remains described by General Relativity, which we never managed to write in a quantum version. It doesn't matter at all in the world of particles, where gravitational forces are extremely weak compared to that of other interactions. At the human scale, both theories can cohabit without communicating: our particles, our atoms obey the Standard Model, whereas our body as a whole obeys gravitation. At the scale of a star, the two theories are still well separated, and influence each other only through global, simple effects, like energy conservation.

In addition, there is a more serious problem: astronomical observations show that the behavior of galaxies, of galaxy clusters, and of the Universe as a whole, do not obey the laws of the Standard Model and General Relativity. It is not the case that this "Standard Model plus General Relativity" pairing appears as wrong at these scales, but it appears at least incomplete. To obtain a good description, it needs to be supplemented with several components, the famous "Dark Matter" and the also famous "Dark Energy". Thanks to accurate astronomical measurements, we are sure that these components do not belong to the present Standard Model. There is a plethora of possible models to describe them, but none of them describes everything, nor do they solve all problems at the same time. No model surpasses others; hence none has reached the status of a new Standard Model.

On one hand we have two theories which ignore each other, and on the other hand we have mysterious observations. Most physicists think that there will be, some day, a common solution to these two problems: a quantum theory for gravitation, and the elucidation of the nature of Dark Matter and Dark Energy. Who knows, the solution might lie in a much larger "change of paradigm", and we — or our grandchildren — will make fun of the naiveté of the 20th century theories.

The ultimate equation?

In the meantime, we use this incomplete and imperfect tool. This is always the case when one is at the forefront of scientific progress. It will require time, hesitations, oscillations, good measurements, wrong measurements that will later be corrected, papers more or less well written, sterile or fruitful intuitions, major or tiny discoveries, in domains apparently as different as cosmology and particle physics at accelerators, so that one day all this jumble of terms could be replaced by a couple of nice characters, symbolizing an elegant and harmonious theory, like the equation at the beginning of this chapter:

$$\boxed{?}\,\Psi = 0$$

Here, $\boldsymbol{\Psi}$ is the fundamental object of nature, the basic component of particles, fields, scalars, spinors, solitons, strings, membranes, curvature, topology…

$\boxed{?}$ stands for the fundamental operator, the unique rule, The Law.

= 0, because this is our way of describing the existence of the World: an object submitted to a law. A trajectory, an equilibrium. Inside "$\boxed{?}\,\Psi$" many things can happen. Energy may be converted into matter, the curvature of space-time may be traded against energy, fluctuations may develop, black holes may grow and evaporate, the Universe itself may be born and disappear (or remain eternal). In short, everything is contained in $\boxed{?}\,\Psi$.

Is this a reality? Does a Universe with this equation as its law *really* exist? Or is this equation only the representation of our time, which our little minds have projected onto the Universe, or even only on themselves?

The question is fascinating, but somewhat vain, since as soon as we reach this this nice intellectual edifice, it will immediately be challenged. Comes a tiny measurement which refuses to enter the framework, or even a thought experiment which points to some internal contradiction, and we will add terms to $\Box\Psi = 0$, we will abandon its purity which would have turned useless, and we will work on:

$$\Box\Psi + \alpha A + \beta B + \gamma C = 0$$

until the next scientific revolution, until the next change of paradigm…

Index

Anderson, 107
Aristotle, 19
Avogadro, 39

Becquerel, 71, 86
Bell, 87
Boltzmann, 39
Boyle, 43
Brout, 124

Copernicus, 19

Dali, 63
Dalton, 39
Darwin, 70
Democritus, 38
Descartes, 11
Dirac, 69, 103, 105–108, 113

Eddington, 71
Einstein, 28, 67, 69, 72, 75, 93, 95
Englert, 124, 127, 131
Euclid, 3
Euler, 12, 13

Fermat, 12, 13
Feynman, 111, 113–116

Galileo, 19
Gell-Mann, 123
Gianotti, 126
Gödel, 88

Hahn, 71
Heisenberg, 86, 88
Higgs, 124, 127, 131
Hooke, 43

Incandela, 126

Jackson, 62

Kelvin, 71
Kepler, 11

Lacan, 77, 116
Lagrange, 12, 13, 30
Laskar, 30
Leibniz, 45
Lemaître, 96
Lévy-Strauss, 116
Lorentz, 69
Loschmidt, 40

Mariotte, 43
Maupertuis, 12

Maxwell, viii, 39, 59, 61
Meitner, 71

Navier, 49
Newton, viii, 19, 25, 51, 88

Oppenheimer, 72

Planck, 69, 75, 77
Poincaré, 69

Roosevelt, 72
Rubbia, 64, 123

Schrödinger, 75, 80
Shakespeare, 38, 63
Snell, 11
Stokes, 49

Van der Meer, 64